COMBINATORIAL DESIGNS AND APPLICATIONS

PURE AND APPLIED MATHEMATICS

A Program of Monographs, Textbooks, and Lecture Notes

LECTURE NOTES

IN PURE AND APPLIED MATHEMATICS

1. *N. Jacobson,* Exceptional Lie Algebras
2. *L. -Å. Lindahl and F. Poulsen,* Thin Sets in Harmonic Analysis
3. *I. Satake,* Classification Theory of Semi-Simple Algebraic Groups
4. *F. Hirzebruch, W. D. Newmann, and S. S. Koh,* Differentiable Manifolds and Quadratic Forms (out of print)
5. *I. Chavel,* Riemannian Symmetric Spaces of Rank One (out of print)
6. *R. B. Burckel,* Characterization of C(X) Among Its Subalgebras
7. *B. R. McDonald, A. R. Magid, and K. C. Smith,* Ring Theory: Proceedings of the Oklahoma Conference
8. *Y.-T. Siu,* Techniques of Extension on Analytic Objects
9. *S. R. Caradus, W. E. Pfaffenberger, and B. Yood,* Calkin Algebras and Algebras of Operators on Banach Spaces
10. *E. O. Roxin, P.-T. Liu, and R. L. Sternberg,* Differential Games and Control Theory
11. *M. Orzech and C. Small,* The Brauer Group of Commutative Rings
12. *S. Thomeier,* Topology and Its Applications
13. *J. M. Lopez and K. A. Ross,* Sidon Sets
14. *W. W. Comfort and S. Negrepontis,* Continuous Pseudometrics
15. *K. McKennon and J. M. Robertson,* Locally Convex Spaces
16. *M. Carmeli and S. Malin,* Representations of the Rotation and Lorentz Groups: An Introduction
17. *G. B. Seligman,* Rational Methods in Lie Algebras
18. *D. G. de Figueiredo,* Functional Analysis: Proceedings of the Brazilian Mathematical Society Symposium
19. *L. Cesari, R. Kannan, and J. D. Schuur,* Nonlinear Functional Analysis and Differential Equations: Proceedings of the Michigan State University Conference
20. *J. J. Schäffer,* Geometry of Spheres in Normed Spaces
21. *K. Yano and M. Kon,* Anti-Invariant Submanifolds
22. *W. V. Vasconcelos,* The Rings of Dimension Two
23. *R. E. Chandler,* Hausdorff Compactifications
24. *S. P. Franklin and B. V. S. Thomas,* Topology: Proceedings of the Memphis State University Conference
25. *S. K. Jain,* Ring Theory: Proceedings of the Ohio University Conference
26. *B. R. McDonald and R. A. Morris,* Ring Theory II: Proceedings of the Second Oklahoma Conference
27. *R. B. Mura and A. Rhemtulla,* Orderable Groups
28. *J. R. Graef,* Stability of Dynamical Systems: Theory and Applications
29. *H.-C. Wang,* Homogeneous Branch Algebras
30. *E. O. Roxin, P.-T. Liu, and R. L. Sternberg,* Differential Games and Control Theory II
31. *R. D. Porter,* Introduction to Fibre Bundles
32. *M. Altman,* Contractors and Contractor Directions Theory and Applications
33. *J. S. Golan,* Decomposition and Dimension in Module Categories
34. *G. Fairweather,* Finite Element Galerkin Methods for Differential Equations
35. *J. D. Sally,* Numbers of Generators of Ideals in Local Rings
36. *S. S. Miller,* Complex Analysis: Proceedings of the S.U.N.Y. Brockport Conference
37. *R. Gordon,* Representation Theory of Algebras: Proceedings of the Philadelphia Conference
38. *M. Goto and F. D. Grosshans,* Semisimple Lie Algebras
39. *A. I. Arruda, N. C. A. da Costa, and R. Chuaqui,* Mathematical Logic: Proceedings of the First Brazilian Conference

40. *F. Van Oystaeyen,* Ring Theory: Proceedings of the 1977 Antwerp Conference
41. *F. Van Oystaeyen and A. Verschoren,* Reflectors and Localization: Application to Sheaf Theory
42. *M. Satyanarayana,* Positively Ordered Semigroups
43. *D. L. Russell,* Mathematics of Finite-Dimensional Control Systems
44. *P.-T. Liu and E. Roxin,* Differential Games and Control Theory III: Proceedings of the Third Kingston Conference, Part A
45. *A. Geramita and J. Seberry,* Orthogonal Designs: Quadratic Forms and Hadamard Matrices
46. *J. Cigler, V. Losert, and P. Michor,* Banach Modules and Functors on Categories of Banach Spaces
47. *P.-T. Liu and J. G. Sutinen,* Control Theory in Mathematical Economics: Proceedings of the Third Kingston Conference, Part B
48. *C. Byrnes,* Partial Differential Equations and Geometry
49. *G. Klambauer,* Problems and Propositions in Analysis
50. *J. Knopfmacher,* Analytic Arithmetic of Algebraic Function Fields
51. *F. Van Oystaeyen,* Ring Theory: Proceedings of the 1978 Antwerp Conference
52. *B. Kedem,* Binary Time Series
53. *J. Barros-Neto and R. A. Artino,* Hypoelliptic Boundary-Value Problems
54. *R. L. Sternberg, A. J. Kalinowski, and J. S. Papadakis,* Nonlinear Partial Differential Equations in Engineering and Applied Science
55. *B. R. McDonald,* Ring Theory and Algebra III: Proceedings of the Third Oklahoma Conference
56. *J. S. Golan,* Structure Sheaves over a Noncommutative Ring
57. *T. V. Narayana, J. G. Williams, and R. M. Mathsen,* Combinatorics, Representation Theory and Statistical Methods in Groups: YOUNG DAY Proceedings
58. *T. A. Burton,* Modeling and Differential Equations in Biology
59. *K. H. Kim and F. W. Roush,* Introduction to Mathematical Consensus Theory
60. *J. Banas and K. Goebel,* Measures of Noncompactness in Banach Spaces
61. *O. A. Nielson,* Direct Integral Theory
62. *J. E. Smith, G. O. Kenny, and R. N. Ball,* Ordered Groups: Proceedings of the Boise State Conference
63. *J. Cronin,* Mathematics of Cell Electrophysiology
64. *J. W. Brewer,* Power Series Over Commutative Rings
65. *P. K. Kamthan and M. Gupta,* Sequence Spaces and Series
66. *T. G. McLaughlin,* Regressive Sets and the Theory of Isols
67. *T. L. Herdman, S. M. Rankin, III, and H. W. Stech,* Integral and Functional Differential Equations
68. *R. Draper,* Commutative Algebra: Analytic Methods
69. *W. G. McKay and J. Patera,* Tables of Dimensions, Indices, and Branching Rules for Representations of Simple Lie Algebras
70. *R. L. Devaney and Z. H. Nitecki,* Classical Mechanics and Dynamical Systems
71. *J. Van Geel,* Places and Valuations in Noncommutative Ring Theory
72. *C. Faith,* Injective Modules and Injective Quotient Rings
73. *A. Fiacco,* Mathematical Programming with Data Perturbations I
74. *P. Schultz, C. Praeger, and R. Sullivan,* Algebraic Structures and Applications Proceedings of the First Western Australian Conference on Algebra
75. *L. Bican, T. Kepka, and P. Nemec,* Rings, Modules, and Preradicals
76. *D. C. Kay and M. Breen,* Convexity and Related Combinatorial Geometry: Proceedings of the Second University of Oklahoma Conference
77. *P. Fletcher and W. F. Lindgren,* Quasi-Uniform Spaces
78. *C.-C Yang,* Factorization Theory of Meromorphic Functions
79. *O. Taussky,* Ternary Quadratic Forms and Norms
80. *S. P. Singh and J. H. Burry,* Nonlinear Analysis and Applications
81. *K. B. Hannsgen, T. L. Herdman, H. W. Stech, and R. L. Wheeler,* Volterra and Functional Differential Equations

82. *N. L. Johnson, M. J. Kallaher, and C. T. Long,* Finite Geometries: Proceedings of a Conference in Honor of T. G. Ostrom
83. *G. I. Zapata,* Functional Analysis, Holomorphy, and Approximation Theory
84. *S. Greco and G. Valla,* Commutative Algebra: Proceedings of the Trento Conference
85. *A. V. Fiacco,* Mathematical Programming with Data Perturbations II
86. *J.-B. Hiriart-Urruty, W. Oettli, and J. Stoer,* Optimization: Theory and Algorithms
87. *A. Figa Talamanca and M. A. Picardello,* Harmonic Analysis on Free Groups
88. *M. Harada,* Factor Categories with Applications to Direct Decomposition of Modules
89. *V. I. Istrăţescu,* Strict Convexity and Complex Strict Convexity: Theory and Applications
90. *V. Lakshmikantham,* Trends in Theory and Practice of Nonlinear Differential Equations
91. *H. L. Manocha and J. B. Srivastava,* Algebra and Its Applications
92. *D. V. Chudnovsky and G. V. Chudnovsky,* Classical and Quantum Models and Arithmetic Problems
93. *J. W. Longley,* Least Squares Computations Using Orthogonalization Methods
94. *L. P. de Alcantara,* Mathematical Logic and Formal Systems
95. *C. E. Aull,* Rings of Continuous Functions
96. *R. Chuaqui,* Analysis, Geometry, and Probability
97. *L. Fuchs and L. Salce,* Modules Over Valuation Domains
98. *P. Fischer and W. R. Smith,* Chaos, Fractals, and Dynamics
99. *W. B. Powell and C. Tsinakis,* Ordered Algebraic Structures
100. *G. M. Rassias and T. M. Rassias,* Differential Geometry, Calculus of Variations, and Their Applications
101. *R.-E. Hoffmann and K. H. Hofmann,* Continuous Lattices and Their Applications
102. *J. H. Lightbourne, III, and S. M. Rankin, III,* Physical Mathematics and Nonlinear Partial Differential Equations
103. *C. A. Baker and L. M. Batten,* Finite Geometries
104. *J. W. Brewer, J. W. Bunce, and F. S. Van Vleck,* Linear Systems Over Commutative Rings
105. *C. McCrory and T. Shifrin,* Geometry and Topology: Manifolds, Varieties, and Knots
106. *D. W. Kueker, E. G. K. Lopez-Escobar, and C. H. Smith,* Mathematical Logic and Theoretical Computer Science
107. *B.-L. Lin and S. Simons,* Nonlinear and Convex Analysis: Proceedings in Honor of Ky Fan
108. *S. J. Lee,* Operator Methods for Optimal Control Problems
109. *V. Lakshmikantham,* Nonlinear Analysis and Applications
110. *S. F. McCormick,* Multigrid Methods: Theory, Applications, and Supercomputing
111. *M. C. Tangora,* Computers in Algebra
112. *D. V. Chudnovsky and G. V. Chudnovsky,* Search Theory: Some Recent Developments
113. *D. V. Chudnovsky and R. D. Jenks,* Computer Algebra
114. *M. C. Tangora,* Computers in Geometry and Topology
115. *P. Nelson, V. Faber, T. A. Manteuffel, D. L. Seth, and A. B. White, Jr.* Transport Theory, Invariant Imbedding, and Integral Equations: Proceedings in Honor of G. M. Wing's 65th Birthday
116. *P. Clément, S. Invernizzi, E. Mitidieri, and I. I. Vrabie,* Semigroup Theory and Applications
117. *J. Vinuesa,* Orthogonal Polynomials and Their Applications: Proceedings of the International Congress
118. *C. M. Dafermos, G. Ladas, and G. Papanicolaou,* Differential Equations: Proceedings of the EQUADIFF Conference
119. *E. O. Roxin,* Modern Optimal Control: A Conference in Honor of Solomon Lefschetz and Joseph P. Lasalle
120. *J. C. Díaz,* Mathematics for Large Scale Computing
121. *P. S. Milojević,* Nonlinear Functional Analysis

122. *C. Sadosky,* Analysis and Partial Differential Equations: A Collection of Papers Dedicated to Mischa Cotlar
123. *R. M. Shortt,* General Topology and Applications: Proceedings of the 1988 Northeast Conference
124. *R. Wong,* Asymptotic and Computational Analysis: Conference in Honor of Frank W. J. Olver's 65th Birthday
125. *D. V. Chudnovsky and R. D. Jenks,* Computers in Mathematics
126. *W. D. Wallis, H. Shen, W. Wei, and L. Zhu,* Combinatorial Designs and Applications

Other Volumes in Preparation

COMBINATORIAL DESIGNS AND APPLICATIONS

Edited by

W. D. WALLIS
Southern Illinois University
Carbondale, Illinois

H. SHEN
Shanghai Jia Tong University
Shanghai, People's Republic of China

W. WEI
Sichuan University
Chengdu, People's Republic of China

L. ZHU
Suzhou University
Suzhou, People's Republic of China

CRC Press
Taylor & Francis Group
Boca Raton London New York

CRC Press is an imprint of the
Taylor & Francis Group, an **informa** business

CRC Press
Taylor & Francis Group
6000 Broken Sound Parkway NW, Suite 300
Boca Raton, FL 33487-2742

First issued in hardback 2017

ISBN 13: 978-1-138-40403-8 (hbk)
ISBN 13: 978-0-8247-8394-5 (pbk)

Visit the Taylor & Francis Web site at
http://www.taylorandfrancis.com

and the CRC Press Web site at
http://www.crcpress.com

Library of Congress Cataloging-in-Publication Data

Combinatorial designs and applications/edited by W.D. Wallis ... [et al.].
p. cm. -- (Lecture notes in pure and applied mathematics: 126)
"[Papers from] the International Conference on Combinatorial Designs and Applications ... held August 18-23, 1988 at ... Hunangshan ... China ... sponsored by the Hefei Branch of Academia Sinica [and others]"--Pref.
Includes bibliographical references and index.
ISBN 0-8247-8394-8 (acid-free paper)
1. Combinatorial designs and configurations--Congresses.
I. Wallis, W. D. II. International Conference on Combinatorial Designs and Applications (1988: Huang-shan shih, Anhwei Province, China) III. Chung-kuo k'o hsüeh yuan. Ho-fei fen yüan.
IV. Series: Lecture notes in pure and applied mathematics; v. 126.
QA166.26.C653 1990
511'.6--dc20 90-3628
CIP

Preface

The International Conference on Combinatorial Designs and Applications was held August 18-23, 1988, at Tunxi (now named Huangshan) City, Anhui Province, China. The conference was sponsored by the Hefei Branch of Academia Sinica, Shanghai Jiao Tong University, Sichuan University, and Suzhou University. The purpose of the conference was to bring together those interested in sharing and listening to current problems and results in the areas of combinatorial designs, their applications, and other related topics. The organizing committee wishes to express its gratitude to the local organizer, the Hefei Branch of Academia Sinica, for administering the conference.

The conference consisted of two parts, forty-minute invited talks and twenty-minute contributed papers. The former consisted of seven talks, given by F. E. Bennett, P. V. Ceccherini, S. M. Dodunekov, H. Shen, W. D. Wallis, W. Wei, and L. Zhu, among which the talks "On minimum matrix representation of closure operations" by Bennett, "Optimal linear codes" by Dodunekov and "Cyclic planar near difference sets of type 1" by Wei are, for various reasons, not included in the Conference

Proceedings. However, an extended abstract of Professor Bennett's presentation has been included. All papers have been refereed, so they do not always represent exactly what was presented, but they provide an accurate picture of the spirit of the conference. Since some Chinese authors write their names with surname first, while others follow the English convention, we have used a mixed style with surnames always in capitals. The editors wish to thank a number of their colleagues who expressed their interest in the conference, but for different reasons were unable to come. We also wish to thank all the participants, the referees and all those who contributed to the success of the conference and to the publication of the Proceedings.

W. D. Wallis
H. Shen
W. Wei
L. Zhu

Contents

Contributors

F. E. BENNETT Mount Saint Vincent University, Halifax, Nova Scotia, Canada

TIANWEN CAI Shanghai Jiao Tong University, Shanghai, People's Republic of China

PIER VITTORIO CECCHERINI Università di Roma "La Sapienza," Rome, Italy

CHANG YAN-XUN Hebei Normal College, People's Republic of China

DU BEILIANG Suzhou University, Suzhou, People's Republic of China

S. GAO Sichuan University, Chengdu, People's Republic of China

B. GONG Suzhou University, Suzhou, People's Republic of China

GUO HAITAO Shanghai Jiao Tong University, Shanghai, People's Republic of China

KANG TAI Chengdu Teachers College, Chengdu, People's Republic of China

KANG QING-DE Hebei Normal College, People's Republic of China

KU TUNG-HSIN Academia Sinica, Hefei Branch, Hefei, Anhui, People's Republic of China

S. MA Shandong University, Jinan, Shandong, People's Republic of China

SHEN HAO Shanghai Jiao Tong University, Shanghai, People's Republic of China

SUN SHIXIN Chengdu Institute of Radio Engineering, Chengdu, People's Republic of China

W. D. WALLIS Southern Illinois University, Carbondale, Illinois

WANG ZHI-JIAN Suzhou Railway Teachers' College, Suzhou, People's Republic of China

R. WEI Suzhou University, Suzhou, People's Republic of China

W-D WEI Sichuan University, Chengdu, People's Republic of China

JULIN WU Qingdao University, Qingdao, People's Republic of China

WU LISHENG Suzhou University, Suzhou, People's Republic of China

MINZHEN WU Shanghai Jiao Tong University, Shanghai, People's Republic of China

M-Y XIA Huazhong Normal University, Wuhan, Hubei, People's Republic of China

QING XIANG Sichuan University, Chengdu, People's Republic of China

BENFU YANG Teacher-Training College of Chengdu, Chengdu, People's Republic of China

YANG ZHAO Dalian University of Technology, Dalian, People's Republic of China

J. YIN Suzhou University, Suzhou, People's Republic of China

ZHANG YUSEN Dalian University of Technology, Dalian, People's Republic of China

ZHU LIE Suzhou University, Suzhou, People's Republic of China

Minimum Matrix Representation of Closure Operations

F. E. BENNETT, Mount Saint Vincent University, Halifax, Nova Scotia, Canada

Lisheng WU, Suzhou University, Suzhou, China

Let b be a column of an m × n matrix M and A a set of its columns. We say that A implies b if and only if M contains no two rows equal in A but different in b. If $\mathcal{L}_M(A)$ denotes the set of columns implied by A, then $\mathcal{L}_M(A)$ is a closure operation and we say that M represents this closure operation. Let $s(\mathcal{L})$ denote the minimum number of rows of the matrices representing a given closure operation $\mathcal{L}$. The k-uniform closure operation on an n-element groundset X is defined by

$$\mathcal{L}_k^n(A) = \begin{cases} X & \text{if} \quad |A| \geq k, \\ A & \text{if} \quad |A| < k. \end{cases}$$

It is known that the closure operation $\mathcal{L}_k^n$ can be represented by some matrix, that is, there is an m × n matrix M such that $\mathcal{L}_M = \mathcal{L}_k^n$. In this paper we shall verify two conjectures of Demetrovics et al. [Discrete Applied Math. 11(1985), 115-128] with regards to the determination of $s(\mathcal{L}_3^n)$.

In particular, we prove that $s(\mathcal{L}_3^n) = n$ for all $n \geq 7$, with the possible exception of $n = 8$.

Existence of Kirkman Systems KS (2,4,v) Containing Kirkman Subsystems

Tianwen CAI, Department of Applied Mathematics, Shanghai Jiao Tong University, Shanghai, China

The obvious necessary conditions for the existence of a Kirkman system KS(2,k,u) containing a sub-KS(2,k,v) are $u \equiv v \equiv k \pmod{k(k-1)}$, $u \geq kv$. In this paper, we consider the embeddings of Kirkman systems KS(2,4,v) and show that there exists a KS(2,4,u) containing a sub-KS(2,4,v) if $u \equiv v \equiv 4 \pmod{12}$, $u \geq 5v-160$, $v \geq 340$. We also get some results for the case $v \leq 328$.

1. INTRODUCTION

A *pairwise balanced design* (PBD) is a pair $(X,\mathcal{A})$ where X is a set of elements (called points) and $\mathcal{A}$ is a set of subsets of X (called blocks) such that every unordered pair of points is contained in a unique block of $\mathcal{A}$. A balanced incomplete block design (BIBD) is a PBD where all blocks have the same size.

We use the notation B(K,v) to denote a PBD$(X,\mathcal{A})$, where $|X| = v$, $|A| \in K$ for every $A \in \mathcal{A}$. When $K = \{k\}$, the PBD is a BIBD, denoted by B(k,v).

Let $(X,\mathcal{A})$ be a PBD. If a set of points $Y \subseteq X$ has the property that for any $A \in \mathcal{A}$, either $|Y \cap A| \leq 1$ or $A \subseteq Y$, then we say that Y is a *subdesign* of the PBD.

A *parallel class* in a PBD is a set of blocks that form a partition of the point set. A PBD is *resolvable* if the block set can be partitioned into parallel classes. A *Kirkman system* KS(2,k,v) is defined to be a resolvable BIBD B(k,v).

We write

$$B^*(k) = \{v \mid \text{there exists a KS}(2,k,v)\}.$$

In [1], Hanani, Ray-Chaudhuri and Wilson showed that the necessary and sufficient condition for the existence of a KS(2,3,v) is $v \equiv 3 \pmod 6$ and there exists a KS(2,4,v) if and only if $v \equiv 4 \pmod{12}$, i.e. $B^*(3) = 6N+3$ and $B^*(4) = 12N+4$.

We say that a Kirkman system KS(2,k,v) is embedded into a Kirkman system KS(2,k,u) (or, a KS(2,k,u) contains a sub-KS(2,k,v)) only if the parallel classes of the KS(2,k,v) are induced by the parallel classes of the KS(2,k,u). The obvious necessary conditions for the existence of a KS(2,k,u) containing a sub-KS(2,k,v) are $u \equiv v \equiv k \pmod{k(k-1)}$, $u \geq kv$.

The embedding problem of Kirkman triple systems KS(2,3,v) was studied by Rees and Stinson in [8], [9], [10] and [11]. They have solved this problem completely. The necessary conditions for the existence of a KS(2,3,u) containing a sub-KS(2,3,v) are $u \equiv v \equiv 3 \pmod 6$, $u \geq 3v$. In [8], [9], [10], and [11], it is shown that these necessary conditions are also sufficient.

In this paper, we are interested in KS(2,4,u) which contains a KS(2,4,v) as a subsystem. We show that for all $u \equiv v \equiv 4 \pmod{12}$, $u \geq 5v-160$, $v \geq 340$, there exists a KS(2,4,u) containing a sub-KS(2,4,v); and for $u \equiv v \equiv 4 \pmod{12}$, $u \geq 4.3v-44$, $v \geq 9340$, there exists a KS(2,4,u) containing a sub-KS(2,4,v).

2. CONSTRUCTIONS FOR KS(2,4,u) CONTAINING SUBSYSTEMS

A *group divisible design* GD(K,M;v) is a triple $(X,\mathcal{G},\mathcal{A})$ which satisfies the following properties:

1) $\mathcal{G}$ is a partition of X into subsets called groups;

2) $\mathcal{A}$ is a set of subsets of X (called blocks) such that a group and a block contain at most one common point;

3) every pair of points from distinct groups occurs in a unique block;

4) $|X| = v$, $|G| \in M$ for every $G \in \mathcal{G}$, and $|A| \in K$ for every $A \in \mathcal{A}$.

A transveral design TD(k,n) is a GD(k,n;kn).

Now we use pairwise balanced designs, group divisible designs and transveral designs to construct KS(2,4,u) containing sub-KS(2,4,v). The following is an important theorem of this paper.

THEOREM 2.1. Suppose $(X,\mathcal{G},\mathcal{A})$ is a GD(K,M;v) satisfying:

1) $k \equiv 1 \pmod 4$, for each $k \in K$ (so that there exists a KS(2,4,3k+1));

2) for every $m \in M$, there exists a KS(2,4,$3m+v_0$) containing a sub-KS(2,4,v_0).

then there exists a KS(2,4,$3v+v_0$) containing a sub-KS(2,4,$3m+v_0$) for each $m \in M$.

PROOF. Let $X = \{x_1,x_2,\ldots,x_v\}$ be the point set of the GD(K,M;v). Give every point weight 3, getting a set $X' = \{x_1,y_1,z_1,x_2,y_2,z_2,\ldots, x_v,y_v,z_v\}$, then consider the groups and blocks respectively.

(i) Let $G = \{x_{i_1},x_{i_2},\ldots,x_{i_m}\} \in \mathcal{G}$. Give every point weight 3 and adjoint v_0 infinite points $e_1,e_2,\ldots,e_{v_0}$. On this $(3m+v_0)$-set, construct a KS(2,4,$3m+v_0$) such that $\{e_1,e_2,\ldots,e_{v_0}\}$ becomes the point set of a sub-KS(2,4,v_0). If we remove the sub-KS(2,4,v_0), we get m parallel classes which are denoted by $P_{i_1},P_{i_2},\ldots,P_{i_m}$ respectively, and $(v_0-1)/3$ partial parallel classes which are denoted by $T_1^G,T_2^G,\ldots,T_{(v_0-1)/3}^G$ respectively.

So, for each group $G \in \mathcal{G}$, we have a KS(2,4,$3m+v_0$) containing a sub-KS(2,4,v_0) of which the point set is $\{e_1,e_2,\ldots,e_{v_0}\}$. We discard all these sub-KS(2,4,v_0) except one of them, and denote the $(v_0-1)/3$ parallel classes of this sub-

$KS(2,4,v_0)$ by $D_1, D_2, \ldots, D_{(v_0-1)/3}$ respectively. Let

$$T_i = \bigcup_{G\in\mathcal{G}} T_i^G, \qquad i = 1,2,\ldots,(v_0-1)/3.$$

(ii) Let $B = \{x_{j_1}, x_{j_2}, \ldots, x_{j_k}\} \in \mathcal{A}$. Give every point weight 3, getting a set $\{x_{j_1}, y_{j_1}, z_{j_1}, x_{j_2}, y_{j_2}, z_{j_2}, \ldots, x_{j_k}, y_{j_k}, z_{j_k}\}$. Then we add one infinite point s to it and construct a $KS(2,4,3k+1)$ on this $(3k+1)$-set such that $\{s, x_{j_1}, y_{j_1}, z_{j_1}\}$, $\{s, x_{j_2}, y_{j_2}, z_{j_2}\}$, ..., $\{s, x_{j_k}, y_{j_k}, z_{j_k}\}$ are its blocks. These k blocks belong to the k parallel classes of the $KS(2,4,3k+1)$. We denote the parallel class which contains the block $\{s, x_{j_r}, y_{j_r}, z_{j_r}\}$ by $H_{j_r}^B$, $r = 1,2,\ldots,k$. Now we discard these k blocks which contain the infinite point s, and denote $F_{j_r}^B = H_{j_r}^B \backslash \{\{s, x_{j_r}, y_{j_r}, z_{j_r}\}\}$, $r = 1,2,\ldots,k$. So, for every block $B \in \mathcal{A}$, we have k holey parallel classes $F_{j_1}^B, F_{j_2}^B, \ldots, F_{j_k}^B$. If $B_1, B_2 \in \mathcal{A}$, and $x_j \in B_1 \cap B_2$, then any two blocks from $F_j^{B_1} \cup F_j^{B_2}$ are disjoint. Let

$$F_j = \bigcup_{\substack{x_j\in B\\ B\in\mathcal{A}}} F_j^B, \qquad j = 1,2,\ldots,v.$$

It is obvious that the blocks we obtain from (i) and (ii) form a $B(4,3v+v_0)$ $(X^*, \mathcal{A}^*)$. The point set X^* is

$$X^* = X' \cup \{e_1, e_2, \ldots, e_{v_0}\}.$$

Now let us point out its parallel classes. $P_1 \cup F_1$, $P_2 \cup F_2$, ..., $P_v \cup F_v$ are v parallel classes and another $(v_0-1)/3$ parallel classes are $D_i \cup T_i$, $i = 1,2,\ldots,(v_0-1)/3$. So, $(X^*, \mathcal{A}^*)$ is resolvable and it is obvious that $(X^*, \mathcal{A}^*)$ contains a sub-$KS(2,4,3m+v_0)$ for each $m \in M$. #

The generalization of Theorem 2.1 is

THEOREM 2.2. Suppose $(X, \mathcal{G}, \mathcal{A})$ is a $GD(K,M;v)$ satisfying:

1) there exists a $KS(2,k,(k-1)k_i+1)$ for each $k_i \in K$;

2) for each $m \in M$, there exists a $KS(2,k,(k-1)m+v_0)$ containing a sub-$KS(2,k,v_0)$.

Then there exists a $KS(2,k,(k-1)v+v_0)$ containing a sub-$KS(2,k,(k-1)m+v_0)$ for each $m \in M$.

Using the relationship between PBD's and GDD's, we have

THEOREM 2.3. If there exists a $B(K,v)$ satisfying:

(i) $k \equiv 1 \pmod 4$ for each $k \in K$;

(ii) for each $k \in K$, there is at least one block of size k.

Then for each $k \in K$, there exists a $KS(2,4,3v+1)$ containing a sub-$KS(2,4,3k+1)$.

PROOF. Deleting one point from the block of size k, we obtain a GDD with block size from $4N+1$, group size from $4N$ and at least one group of size $k-1$. The claim follows from Theorem 2.1 with $v_0 = 4$. #

The following result is due to R.D. Baker [2]:

THEOREM 2.4. Suppose there exists a $TD(17,n)$. Then for all $0 \le m_1, m_2 \le n$, there exists a $GD(\{5,17\},\{n,n+4m_1,n+4m_2\}, 17n+4(m_1+m_2))$.

This theorem furnishes us a series of GDD's which are useful in construction of a $KS(2,4,u)$ containing a sub-$KS(2,4,v)$. By Theorem 2.1 and Theorem 2.4, we have

THEOREM 2.5. Suppose there exists a $TD(17,n)$, $0 \le m_1, m_2 \le n$. Suppose there exist a $KS(2,4,3n+12m_i+v_0)$ $(i = 1,2)$ and a $KS(2,4,3n+v_0)$ such that each of them contains a sub-$KS(2,4,v_0)$. Then there exists a $KS(2,4,51n+12(m_1+m_2)+v_0)$ containing a sub-$KS(2,4,3n+v_0)$ and a sub-$KS(2,4,3n+12m_i+v_0)$ $(i = 1,2)$.

COROLLARY 2.6. If there exists a $TD(17,n)$, $n \in 4N+1$, then for all $0 \le m_1, m_2 \le n$, there exists a $KS(2,4,51n+12(m_1+m_2)+1)$ containing a sub-$KS(2,4,3n+12m_i+1)$ $(i = 1,2)$.

PROOF. This is the case $v_0 = 1$ of Theorem 2.5. #

COROLLARY 2.7. If there exists a $TD(17,n)$, $n \in 4N$, then for all $0 \leq m_1, m_2 \leq n$, there exists a $KS(2,4,51n+12(m_1+m_2)+4)$ containing a sub-$KS(2,4,3n+12m_i+4)$ $(i = 1,2)$.

PROOF. This is the case $v_0 = 4$ of Theorem 2.5. #

Now we introduce some notations which shall be useful in this paper. Let

$$A = \{n \mid N(n) \geq 15,\ n \in 4N+1,\ n \geq 25\}.$$

Then there exists a $TD(17,n)$ for each $n \in A$. Let $n \in A$. We denote the smallest integer of A which is larger than n by $n' = n+4\tau(n)$, i.e.

$$\tau(n) = \frac{1}{4}\{\min\ (n'-n):\ n' > n,\ n' \in A\}$$

LEMMA 2.8. (i) $n > \tau(n)$, for all $n \in A$;
(ii) $n \geq 16\tau(n)+9$, for $37 \leq n \in A$;
(iii) $\tau(n) \leq 5$, for $641 \leq n \in A$.

PROOF. Brouwer and Rees [4] showed that $N(n) \geq 15$ for $n \geq 10632$, and $o_{15} \leq 3603$, where $o_{15} = \max\ \{v \mid v \text{ odd and } N(v) < 15\}$. So we just need to verify the case $n \leq 3603$, and the claim easily follows from [3]. #

Apply Lemma 2.8 and using induction, we have

LEMMA 2.9. Suppose $v \in B^*(4)$, $v \geq 76$, then there exist $n \in A$ and $0 \leq m \leq n$, such that $v = 3n+12m+1$.

LEMMA 2.10. Suppose $v \in B^*(4)$, $v \geq 568$, then there exists $n \in A$ such that $15n+13 \leq v \leq 15(n+4\tau(n))+1$.

By Corollary 2.6 and Lemma 2.9, we obtain

THEOREM 2.11. Let $v \in B^*(4)$, $v \geq 76$. Suppose $n \in A$, $0 \leq m \leq n$ such that $v = 3n+12m+1$, then for $u \in B^*(4)$, $51n+12m+1 \leq u \leq 63n+12m+1$, there exists a $KS(2,4,u)$ containing a sub-$KS(2,4,v)$.

LEMMA 2.12. If $v \in B^*(4)$, then there exists a $KS(2,4,4v)$ containing a sub-$KS(2,4,v)$.

PROOF. By [6] Remark 3.5, there exists a $B(\{5,4t+1^*\},16t+5)$ for each $t \in N$ (where the notation $4t+1^*$ means that there is just one block of size $4t+1$). Applying Theorem 2.3, we obtain a $KS(2,4,48t+16)$ containing a sub-$KS(2,4,12t+4)$ for each $t \in N$. #

It follows from Lemma 2.12 directly that there exists a $KS(2,4,4^n v)$ containing a sub-$KS(2,4,v)$ for each $n \in N$.

THEOREM 2.13. Let $v \in B^*(4)$, $v \geq 76$. Suppose $25 \leq n_0 \in A$, $0 \leq m_0 \leq n_0$ such that $v = 3n_0+12m_0+1$. Then for all $u \in B^*(4)$, $u \geq 51n_0+12m_0+1$, there exists a $KS(2,4,u)$ containing a sub-$KS(2,4,v)$.

PROOF. The existence of n_0 follows from Lemma 2.9.

$$v = 3n_0+12m_0+1.$$

By Theorem 2.11, there exists a $KS(2,4,u)$ containing a sub-$KS(2,4,v)$ for $u \in B^*(4)$, $51n_0+12m_0+1 \leq u \leq 63n_0+12m_0+1$. If we can prove that there exists $n^* \in A$, $0 \leq m_1^* < m_2^* \leq n^*$ satisfying:

(a) $v = 3n^*+12m_1^*+1$, and for $u \in B^*(4)$, $51n_0+12m_0+1 \leq u \leq 63n^*+12m_1^*+1$, there exists a $KS(2,4,u)$ containing a sub-$KS(2,4,v)$,

(b) $4v = 3n^*+12m_2^*+1$, so, for $51n^*+12m_2^*+1 \leq u \leq 63n^*+12m_2^*+1$, $u \in B^*(4)$, there exists a $KS(2,4,u)$ containing a sub-$KS(2,4,4v)$.

Then applying the same idea for $4^n v$ $(n = 1,2,\ldots)$, we can obtain the conclusion by Lemma 2.12. There are two cases depending on m_0.

Case 1. $m_0 < \tau(n_0)$. Let $n_0 = 4d+1$.

$$4v = 12n_0+48m_0+4 = 3n_0+12(4m_0+3d+1)+1.$$

By Lemma 2.8, we have $\tau(n_0) < 4m_0+3d+1 \leq n_0$ for $n_0 \geq 37$. When $n_0 = 25, 29$, $\tau(25) = 1$ and $\tau(29) \leq 2$, so the inequality also holds. Take $n^* = n_0$, $m_1^* = m_0$, $m_2^* = 4m_0+3d+1$.

Case 2. $m_0 \geq \tau(n_0)$. Put $n_1 = n_0+4\tau(n_0) \in A$, $m_1 = m_0-\tau(n_0)$, then

$$v = 3n_0+12m_0+1 = 3n_1+12m_1+1.$$

It follows from Theorem 2.11 that there exists a KS(2,4,u) containing a sub-KS(2,4,v) for $u \in B^*(4)$, $51n_1+12m_1+1 \leq u \leq 63n_1+12m_1+1$. By Lemma 2.8, $n \geq 16\tau(n)+9$ for $n \geq 37$ and for $n = 25$, so $51n_1+12m_1+1 \leq 63n_0+12m_0+1 \leq 63n_1+12m_1+1$ for $n_0 = 25$ or $n_0 \geq 37$. Though we do not know whether $n_0 \geq 16\tau(n_0)+9$ for $n_0 = 29$, we can take $n_1' = 32$, and a similar result comes from Corollary 2.7. So we can take n_0 from 25 up.

If $m_1 < \tau(n_1)$, we turn to case 1. If $m_1 \geq \tau(n_1)$, using the same method, we get $n_2 = n_1+4\tau(n_1)$, $m_2 = m_1-\tau(n_1)$, and for $u \in B^*(4)$, $51n_0+12m_0+1 \leq u \leq 63n_2+12m_2+1$, there exists a KS(2,4,u) containing a sub-KS(2,4,v). If $m_2 < \tau(n_2)$, we turn to the case 1. If $m_2 \geq \tau(n_2)$, we continue in the same way. Since $m_0 > m_1 > m_2 > \ldots \geq 0$, we can obtain an $m_s < \tau(n_s)$ such that $v = 3n_s+12m_s+1$ and for $u \in B^*(4)$, $51n_0+12m_0+1 \leq u \leq 63n_s+12m_s+1$, there exists a KS(2,4,u) containing a sub-KS(2,4,v). Then we turn to case 1. Now the claim holds. #

3. MAIN RESULTS

With the preparation of section 2, now we can prove

THEOREM 3.1. If $u,v \in B^*(4)$, $u \geq 5v-160$, $v \geq 340$, then there exists a KS(2,4,u) containing a sub-KS(2,4,v).

PROOF. We prove the theorem in three parts.

(1) For $v \geq 568$, $v \in B^*(4)$, according to Lemma 2.10, there exists $37 \leq n \in A$ and $1 \leq t \leq 5\tau(n)$ such that $v = 15n+12t+1 = 3(n+4\tau(n))+12(n-\tau(n)+t)+1$. It follows from Theorem 2.13 that there exists a KS(2,4,u) containing a sub-KS(2,4,v) for $u \in B^*(4)$, $u \geq 63n+192\tau(n)+12t+1$. By Lemma 2.8, we have

$$5v-160 = 75n+60t-155 \geq 63n+192\tau(n)+12t+1, \quad 37 \leq n \in A.$$

Therefore, when $u,v \in B^*(4)$, $u \geq 5v-160$, $v \geq 568$, there exists a

KS(2,4,u) containing a sub-KS(2,4,v).

(2) For $484 \le v \le 556$, $v \in B^*(4)$, we take $n = 37$, $v = 3\times37+12m+1$, $31 \le m \le 37$. By Theorem 2.13, there exists a KS(2,4,u) containing a sub-KS(2,4,v) for $u \in B^*(4)$, $u \ge 51\times37+12m+1$.

$$5v-160 = 15\times37+60m-155 \ge 51\times37+12m+1, \quad 31 \le m \le 37$$

So, there exists a KS(2,4,u) containing a sub-KS(2,4,v) for $u,v \in B^*(4)$, $u \ge 5v-160$, $484 \le v \le 556$.

(3) For $424 \le v \le 472$, $388 \le v \le 412$, $340 \le v \le 376$, we take $n = 32$, $n = 29$, and $n = 25$ respectively. Using a similar method, we obtain that there exists a KS(2,4,u) containing a sub-KS(2,4,v) for $u,v \in B^*(4)$, $u \ge 5v-160$, $340 \le v \le 472$. #

The following is a slight improvement of this theorem for large v.

THEOREM 3.2. If $u,v \in B^*(4)$, $u \ge 4.3v-44$, $v \ge 9340$, then there exists a KS(2,4,u) containing a sub-KS(2,4,v).

PROOF. By Lemma 2.10, for $v \in B^*(4)$, $v \ge 568$, there exists $n \in A$, $1 \le t \le 5\tau(n)$ such that

$$v = 15n+12t+1.$$

So, for $u \in B^*(4)$, $u \ge 63n+192\tau(n)+12t+1$, there exists a KS(2,4,u) containing a sub-KS(2,4,v). By Lemma 2.8, we have $641 \le n \in A$.

$$4.3v-44 = 4.3(15n+12t+1)-44 \ge 63n+192\tau(n)+12t+1$$

Therefore, there exists a KS(2,4,u) containing a sub-KS(2,4,v) for $u,v \in B^*(4)$, $u \ge 4.3v-44$, $v \ge 9628$. For $9340 \le v \le 9616$, we take $n = 641$.

$$v = 3\times641+12m+1, \quad 618 \le m \le 641.$$

It follows from Theorem 2.13 that there exists a KS(2,4,u) containing a sub-KS(2,4,v) for $u \in B^*(4)$, $u \ge 51\times641+12m+1$.

$$4.3v-44 = 4.3(3\times641+12m+1)-44 \ge 51\times641+12m+1$$

for $618 \le m \le 641$. So, there exists a KS(2,4,u) containing a sub-KS(2,4,v) for $u,v \in B^*(4)$, $u \ge 4.3v-44$, $9340 \le v \le 9616$. #

THEOREM 3.3. If $u,v \in B^*(4)$, $u \ge 17v-16$, $v \ge 76$, then there exists a KS(2,4,u) containing a sub-KS(2,4,v).

PROOF. If $v \ge 340$, the claim comes from Theorem 3.1. For $76 \le v \le 328$, take n from 25 up, we obtain the following result:

There exists a KS(2,4,u) containing a sub-KS(2,4,76+12t) for $u \in B^*(4)$, $u \ge 1276+12t$ ($t = 0,1,\ldots,21$).

$$17(76+12t)-16 \ge 1276+12t$$

So, the claim holds. #

By Lemma 2.12, we have

THEOREM 3.4. (1) For $u \in B^*(4)$, $u \ge 1456$, there exists a KS(2,4,u) containing a sub-KS(2,4,64) and a sub-KS(2,4,16);

(2) For $u \in B^*(4)$, $u \ge 1408$, there exists a KS(2,4,u) containing a sub-KS(2,4,52);

(3) For $u \in B^*(4)$, $u \ge 1360$, there exists a KS(2,4,u) containing a sub-KS(2,4,40);

(4) For $u \in B^*(4)$, $u \ge 1312$, there exists a KS(2,4,u) containing a sub-KS(2,4,28).

Now let us consider the case $v = 16$. By Theorem 2.3, we have

LEMMA 3.5. If there exists a $B(\{5,4t+1\},v)$ with at least one block of size five, then there exists a KS(2,4,3v+1) containing a sub-KS(2,4,16).

LEMMA 3.6. (i) There exists a B(5,v) if and only if $v \equiv 1,5 \pmod{20}$;

(ii) Suppose $v \in B^*(5)$, $1 \le t \le (v-1)/4$. Then there exists a $B(\{5,4t+1^*\}, 4v+4t+1)$;

(iii) Suppose there exists a TD(6,g) and $g \equiv 0,\pm 1 \pmod 5$,

$1 \le e \le g$, then there exists a $B(\{5,4e+1^*\},20g+4e+1)$.

LEMMA 3.7 (Ying [5]). (i) If $v \equiv 9,17 \pmod{20}$, $v \notin E$, then there exists a $B(\{5,9^*\},v)$. $E = \{17,29,49,57,69,77,97,117,129,137,157,169,197,277,397,449,497,557,637,717,749,777,797,897\}$.

(ii) If $0 \le a \le 3t+1$, $t > 1$, then there exists a $B(\{5,4t+4a+1^*\}, 64t+4a+21)$.

LEMMA 3.8. There exist $B(\{5,13,17\},77)$, $B(\{5,17\},97)$, $B(\{5,29\},129)$ and $B(\{5,13\},333)$.

The following result comes from the above Lemmas.

THEOREM 3.9. Suppose $u \in B^*(4)$, $u \ge 64$ and $u \notin \{88,100,148,172,220,280,412\}$. Then there exists a $KS(2,4,u)$ containing a sub-$KS(2,4,16)$.

Below we list the results for $v \le 328$ which follow from the above. The numbers in column "O.S" are the order of subsystems, and the numbers in column "O.L" are the order of larger systems which contain the corresponding subsystem.

O.S	O.L	O.S	O.L
328	$u \in B^*(4)$, $u \ge 1528$	196	$u \in B^*(4)$, $u \ge 1396$
316	$u \in B^*(4)$, $u \ge 1516$	184	$u \in B^*(4)$, $u \ge 1384$
304	$u \in B^*(4)$, $u \ge 1504$	172	$u \in B^*(4)$, $u \ge 1372$
292	$u \in B^*(4)$, $u \ge 1492$	160	$u \in B^*(4)$, $u \ge 1360$
280	$u \in B^*(4)$, $u \ge 1480$	148	$u \in B^*(4)$, $u \ge 1348$
268	$u \in B^*(4)$, $u \ge 1468$	136	$u \in B^*(4)$, $u \ge 1336$
256	$u \in B^*(4)$, $u \ge 1456$	124	$u \in B^*(4)$, $u \ge 1324$
244	$u \in B^*(4)$, $u \ge 1444$	112	$u \in B^*(4)$, $u \ge 1312$
232	$u \in B^*(4)$, $u \ge 1432$	100	$u \in B^*(4)$, $u \ge 1300$
220	$u \in B^*(4)$, $u \ge 1420$	88	$u \in B^*(4)$, $u \ge 1288$
208	$u \in B^*(4)$, $u \ge 1408$	76	$u \in B^*(4)$, $u \ge 1276$
64	$u \in B^*(4)$, $u \ge 1456$	28	$u \in B^*(4)$, $u \ge 1312$
52	$u \in B^*(4)$, $u \ge 1408$	16	$u \in B^*(4)$, $u \ge 64$, $u \ne 88,100,148,172,220,280,412$
40	$u \in B^*(4)$, $u \ge 1360$		

ACKNOWLEDGEMENT

The author is indebted to Professor Shen Hao for his helpful advice, encouragement, and many stimulating discussions.

REFERENCES

[1] H. Hanani, D.K. Ray-Chaudhuri and R.M. Wilson, On resolvable designs, Discrete Math. 3(1972), 75-97.

[2] R.D. Baker, Whist tournaments, Congressus Num. 14, 89-100.

[3] A.E. Brouwer, The number of mutually orthogonal Latin squares Math Centrum Amsterdam, Report/ZW123/79 (1979).

[4] A.E. Brouwer and G.H.J. Van Rees, More mutually orthogonal La squares, Discrete Math. 39(1982), 263-281.

[5] J. Ying, On (2,5,v) packing designs, preprint.

[6] R.M. Wilson, Constructions and uses of pairwise balanced designs, Math. Centre Tracts 55(1974), 18-41.

[7] T. Beth, D. Jungnickel and H. Lenz, Design theory, Bibliographisches Institut, Zurich, 1985.

[8] R. Rees and D.R. Stinson, Kirkman triple systems with maximum subsystems, Ars Combinatoria 25(1988), 125-132.

[9] D.R. Stinson, Frames for Kirkman triple systems, Discrete Mat 65(1987), 289-300.

[10] R. Rees and D.R. Stinson, On combinatorial designs with subdesigns, Discrete Math. 77(1989), 259-279.

[11] R. Rees and D.R. Stinson, On the existence of Kirkman triple systems containing Kirkman subsystems, Ars Combinatoria 26(1988), 3-16.

Studying Structures by Counting Geodesics

Pier Vittorio CECCHERINI, Dipartimento di Matematica "G. Castelnuovo", Università di Roma "La Sapienza", Città Universitaria, 00185 Roma, Italy

Hypercubes and graphs associated with vector spaces over a finite field can be characterized by arithmetical properties concerning geodesics (shortest paths). New proofs and some improvements of known results in this direction are here presented. Moreover the case of a product graph is also considered.

1. INTRODUCTION

All sets will be *finite* and all graphs will be *simple*, i.e. without loops or multiple edges. If G = (V,E) is a connected graph, the *distance* d(x,y) of two distinct vertices x and y is the *length* (= number of edges) of a *geodesic* g(x,y), i.e. of a shortest path between x and y; the cardinality of the set $\Gamma(x,y)$ of all the geodesics g(x,y) will be denoted by $\gamma(x,y)$. For any vertex x, it is convenient to assume d(x,x) = 0 and $\gamma(x,x) = 1$.

We shall present the following method of studying structures. Suppose we have a certain finite structure S, e.g. a finite vector space. We want:

-to define a graph G(S) = (V,E) associated with S;

-to find necessary conditions for G(S) involving the numbers $\gamma(x,y)$;

-given any graph G satisfying the above arithmetical conditions, to discuss the problem of reconstructing some structure S such that $G \cong G(S)$.

The following examples will be particularly considered: (a) S is the boolean lattice of all the subsets of a finite set (Thm. 2.1 and Thm. 2.2); (b) S is a finite projective space (Thm. 3.3 and Thm. 3.4); (c) S is a finite vector space (Cor. 3.5).

Case (a) is known as Foldes' characterization of hypercubes (given in [5]); cases (b) and (c) are the *q-analogue* of case (a), cf. [1]. All proofs are here included, because they are much simpler than in the original papers [1] and [5]; the statement of Cor. 3.9 is new. Actually an unnecessary hypothesis (namely that every vertex of G belongs to a fixed geodetic g(m,M)) is removed from the original Thm. 4.1 of [1], cf. Thm. 3.3 and Thm. 3.4; as a consequence we obtain that now, for q = 1, these general theorems reduce *precisely* (and not only in a slightly different form) to the Foldes' characterization of hypercubes.

The previous combinatorial approach suggests studying arithmetical properties of the set of geodesics between two vertices of a given graph G. In section 4 we discuss the case when G is a product graph; this case includes hypercubes: $G = (K_2)^n$, and Hamming graphs: $G = (K_a)\times\ldots\times(K_b)$ with $a,b \geq 2$.

2. FOLDES' CHARACTERIZATION OF HYPERCUBES

We present a simplified proof of a theorem of S. Foldes [5], according to which a (simple) graph is a hypercube iff it is connected, bipartite and the number of geodesics between any two vertices is the factorial of their distance.

If X is a set, denote by B(X) the boolean lattice of all subsets of X, and by $B_k(X)$ the set of all subsets of X with k elements. We shall define the *hypercube on X*, denoted by Q(X), to be the undirected graph G = (V,E) with vertex set V = V(G) = B(X) and having as set E = E(G) of edges the covering relation on $B(X)$ (i.e., $(x,y) \in E$ iff the subset x of X is obtained from the subset y of X by adding or by deleting exactly one element of X).

If n is a natural number, the *hypercube of dimension n* (or *n-cube*), denoted by Q_n, is defined (uniquely up to isomorphisms) as the hypercube $Q(X)$ on a set X with $|X| = n$ elements (for fixing the ideas, assume $X = V(Q_n)$ to be the set of the first n natural numbers).

2.1 THEOREM. ([5]). For any natural number n, the n-cube Q_n is a simple graph $G = (V,E)$ satisfying the following properties:

G_1) G is connected and bipartite;

G_2) For any two vertices $x,y \in V$ the number of geodesics $g(x,y)$ is $\gamma(x,y) = d(x,y)!$;

G_3) $\operatorname{diam} G = n$.

PROOF. G_1) If $(x,y) \in E(Q_n)$ is an edge, then $|x| = |y| \pm 1$. So Q_n is bipartite, with $V(Q_n) = A \cup B$, where A (resp. B) is the set of vertices of odd (resp. even) cardinality. Moreover, Q_n is connected; given any two vertices x and y, a (shortest) path $g(x,y)$ from x to y can be obtained by deleting successively each element of x which does not belong to y and then by adding each element of y which does not belong to x.

G_2) The length of such a geodesic $g(x,y)$ is the distance $d(x,y)$, and it is given by the cardinality of the symmetric difference $x \oplus y$. The number $\gamma(x,y)$ of geodesics from x to y is $d(x,y)!$ because it is the number of permutations on the set $x \oplus y$ of elements which must be deleted or added.

G_3) $\operatorname{diam}(Q_n) = \max\{|x \oplus y| : x,y \subseteq X\} = d(\emptyset,X) = n$. □

2.2 THEOREM ([5]). Let $G = (V,E)$ be a simple graph satisfying conditions G_1-G_3. Then $G \cong Q_n$.

PROOF. Fix a diametral vertex $m \in V$ and consider the partition

$$V = V_0 \cup V_1 \cup \ldots \cup V_n, \quad V_k = \{x \in V: d(m,x) = k\},$$
$$k = 0,1,\ldots,n.$$

As G is bipartite: $(x,y) \in E \Rightarrow \exists!\ k \in \{0,1,\ldots,n\}$: $x \in V_k$ and $y \in V_{k\pm1}$. We shall prove that $G \cong Q_n$ and that $n = |V_1|$. Define a map $^*: V \to B(V_1)$ by assuming:

$$(x \in V) \qquad x^* := \{x_1 \in V_1 : x_1 \text{ belongs to some geodesic } g(m,x)\}.$$

Note that $x_k \in V_k \Rightarrow |x_k^*| = k$, because when x_1 describes all the elements of x_k^* we get

$$\gamma(m,x_k) = \gamma(m,x_1)\cdot|x_k^*|\cdot\gamma(x_1,x_k) \Rightarrow |x_k^*| = k!/(1!(k-1)!) = k.$$

We shall prove, by induction on $k = 0,1,\ldots,|V_1|$, that conversely,

(2.1) For each subset X of V_1 with $|X| = k$, there exists exactly one element $x_k \in V_k$ such that $X = x_k^*$.

This is trivial for $k = 0$ or $k = 1$. The case $k = 2$ is instructive. Let $X = \{x_1,y_1\}$. Then $d(x_1,y_1) = 2$, because $g(x_1,y_1) := (x_1,m,y_1)$ is a geodesic. As $\gamma(x_1,y_1) = 2! = 2$, then there exists exactly one element $x_2 \in V_2$ such that $g'(x_1,y_1) := (x_1,x_2,y_1)$ is the other geodesic between x_1 and y_1; so $X = x_2^*$. The induction hypothesis is that, for each subset Y of V_1 with $|Y| = k-1$, there exists exactly one element $x_{k-1} \in V_{k-1}$ such that $Y = x_{k-1}^*$, $2 \le k-1 < |V_1|$. Let X be a subset of V_1 with $|X| = k$, let x_1 be a *fixed* element of X and put $Y = X\setminus\{x_1\}$; then, by induction hypothesis, $Y = x_{k-1}^*$ for exactly one $x_{k-1} \in V_{k-1}$.

Actually $d(x_1,x_{k-1}) > k-1$ because $x_1 \notin Y$ and G has no odd cycles; if y_1 is an element of Y, a geodesic $g(y_1,x_{k-1})$ has length $k-2$, so that $(x_1,m,g(y_1,x_{k-1}))$ is a path $p(x_1,x_{k-1})$ of length k, and so $d(x_1,x_{k-1}) = k$ and $\gamma(x_1,x_{k-1}) = k!$.

The vertex x_{k-1} is adjacent with $\alpha = k-1$ vertices $x_{k-2} \in V_{k-2}$, because $\gamma(m,x_{k-1}) = \alpha\ \gamma(m,x_{k-2})$ implies $\alpha = (k-1)!/(k-2)!$; for such an $x_{k-2} \in V_{k-2}$ we have $d(x_1,x_{k-2}) = k-1$, so that each geodesic $g(x_1,x_{k-2})$ gives one geodesic $g(x_1,x_{k-1})$. In this way we get $(k-1)\gamma(x_1,x_{k-2}) = (k-1)!(k-1)$ geodesics $g(x_1,x_{k-1})$ having a final edge of the type $(x_{k-2},x_{k-1}) \in E$. The remaining

$$\gamma(x_1,x_{k-1})-(k-1)!(k-1) = k!-(k-1)!(k-1) = (k-1)!$$

other geodesics $g(x_1,x_{k-1})$ must have a final edge of the type (x_k,x_{k-1}) with $x_k \in V_k$. It follows that there exists at least one such an $x_k \in V_k$; for this x_k obviously we have $x_k^* = Y \cup \{x_1\} = X$; moreover such an $x_k \in V_k$ is necessarily unique, because x_k gives all the remaining $(k-1)!$ geodesics since $d(x_1,x_k) = k-1$. So (2.1) is proved. As a consequence we get that V_n is a singleton, i.e. there exists exactly one vertex x_n at maximal distance n from m, and $x_n^* = V_1$, $|V_1| = |x_n^*| = n$.

The bijection $^*: V \to B(V_1)$ gives an isomorphism $G \to Q_n = Q(V_1)$ of graphs, because $(x_{k-1},x_k) \in E(G)$ iff $(x_{k-1}^*, x_k^*) \in E(Q_n)$, as proved just before. □

3. A q-ANALOGUE OF THE PREVIOUS CHARACTERIZATION OF HYPERCUBES

The q-analogue of the boolean lattice $B(X)$ of all the subsets of a finite set X with $n \geq 0$ elements is the lattice $F(P(n-1,q))$ of all the flats of a generalized projective space $P(n-1,q)$ of dimension n-1 and of order $q \geq 1$, cf. Theorem 3.1(a). We recall that a *generalized projective space* of *dimension* n-1 and of *order* $q \geq 1$, $P(n-1,q)$, can be defined by the following three axioms.

P_1) $P(n-1,q)$ is a *linear space of order* q. In other words, $P(n-1,q)$ is a pair $(\mathbf{P},\mathbf{L})$, where $\mathbf{P}$ is a set of elements (called *points*) and $\mathbf{L}$ is a set of subsets of $\mathbf{P}$ (called *lines*) such that every two distinct points x and y belong to exactly one line (called their *joining line* and denoted by $x \vee y$), and every line has q+1 points.

A *flat* of $P(n-1,q)$ is then defined as any subset Y of $\mathbf{P}$ containing the line joining any two distinct points of Y. It follows that the empty set $\emptyset$, each point, each line and $\mathbf{P}$ itself are flats. We denote by $F(P(n-1,q))$ the $(\cap,\vee)$-lattice of all the flats of $P(n-1,q)$.

P_2) The *linear closure* π of any three non-collinear points x,y,z of $\mathbf{P}$ is a *generalized projective plane of order* q. In other words, the minimum flat π containing x,y,z is a linear space of order q such that any two distinct lines of π intersect in one point.

P_3) There is a maximal chain of flats from $\emptyset$ to P of length n. (It follows that every such chain has length n).

Any flat x of a P(n-1,q) turns out to be a P(k-1,q), for a suitable k, with $0 \leq k \leq n$. We shall denote by $F_{k-1}(P(n-1,q))$ the set of all (k-1)-dimensional flats of P(n-1,q). It is easy to prove the following theorem.

3.1 THEOREM (cf. [8]). All the generalized projective spaces P(n-1,q) of dimension $n-1 \geq 1$ and of order $q \geq 1$ are the following.

(a) A generalized projective space of dimension n-1 and of order q = 1 is a boolean lattice over a set X with n elements, and conversely: $F(P(n-1,1)) = B(X)$ and $F_{k-1}(P(n-1,q)) = B_k(X)$, $k = 0,1,\ldots,n$.

(b) A generalized projective space of dimension n-1 and of order $q \geq 2$ is a projective space of dimension n-1 and of order q, and conversely. More precisely:

(b1) A space of dimension 1 is a line; in other words, $F(P(1,q)) = \{Y \in B(X) : |Y| \in \{0,1,q+1\}\}$, where X is a given set with q+1 elements.

(b2) Spaces of dimension 2 are planes: any P(2,q) with $q \geq 2$ is a projective plane of order q, either the desarguesian one (i.e. the projective plane PG(2,q) over the finite field GF(q) with q elements) or a non-desarguesian projective plane of order q.

(b3) A space of dimension $n-1 \geq 3$ and of order $q \geq 2$ is a (necessarily desarguesian) projective space of the same dimension and of the same order, i.e. P(n-1,q) is necessarily the projective space PG(n-1,q) of dimension n-1 over GF(q). □

From (a) of Theorem 3.1, it follows that, whenever P(n-1,q) is a given generalized projective space of dimension n and of order $q \geq 1$, we can assume as a q-analogue of the hypercube Q_n the graph G defined by $G := Q_{n,q} := G(P(n-1,q)) := (V,E)$, where the set V of vertices is given by the set F(P(n-1,q)) of all the flats of the space P(n-1,q) and the set E of edges is the covering relation in the lattice F(P(n-1,q)); in other words, $(x,y) \in E$ iff one of the flats x and y is contained in the other and dim x = dim y±1. For q = 1, $Q_{n,q}$ reduces to $Q_{n,1} = Q_n$, the classical n-cube.

3.2 DUALITY. In a generalized projective space P(n-1,q), a *graphic theorem* is a theorem $T = T(X_{h-1},\dots,X_{k-1}; \subseteq,\cap,\vee)$ - concerning flats of given dimensions h-1,...,k-1, their inclusion, intersections and joinings - which is a consequence of axioms P_1-P_3. By *duality* we can change T into the *dual theorem* $T^* = T(Y_{n-h-1},\dots,Y_{n-k-1}: \supseteq,\vee,\cap)$. T^* is obtained from T by changing each (t-1)-flat of P(n-1,q) in a (n-t-1)-flat of P(n-1,q) and changing $\cap$ with $\vee$ and reversing inclusion. In particular, when P(n-1,q) is a projective space, we have the *duality of projective geometry*, cf. [8].

3.3 NOTATION. Given any integers $q \geq 1$ and $1 \leq k \leq n$, write [ ] for $[\ ]_q$ and define:

$$[0] := 0,\ [1] := 1,\ [k] := q^{k-1}+q^{k-2}+\dots+q+1$$
(= the number of points of P(k-1,q)),
$$[0]! = [1]! = 1,\quad [k]! = [k][k-1]\dots[1],$$
$$C(n,0;q) := 1,\quad C(n,k;q) := [n]!/([k]![n-k]!)$$
$$= [n]\dots[n-k+1]/([k]\dots[1])$$
(= the number of (k-1)-flats of a P(n-1,q))

3.4 THEOREM (cf. [1]). Given any generalized projective space P(n-1,q), the hypercube $Q_{n,q}$ is a simple graph G = (V,E) satisfying the following properties:

G_1) G is connected and bipartite;

$G_2)_q$ There exists a vertex m with the property that for any two vertices $x,y \in V$ the number $\gamma(x,y)$ of geodesics g(x,y) is

$$\gamma(x,y) = d(x,y)![\delta_1(x,y)]![\delta_2(x,y)]!/(\delta_1(x,y)!\delta_2(x,y)!) \tag{3.1}$$

where

(3.2) $\delta_{1,2}(x,y) = (d(x,y) \mp d(m,x) \pm d(m,y))/2,$ and $[\ ] = [\ ]_q;$

$G_3)$ diam $G = n$.

PROOF. $G_1)$ The graph $Q_{n,q}$ associated to $P(n-1,q)$ is obviously a simple connected graph, which is bipartite by the parity of dimensions, because if (x,y) is an edge, then dim x = dim $y \pm 1$.

$G_2)_q$ When dim $x = k-1$, we shall write more carefully $x = x_k$; note that k is the length of any maximal chain of flats from $\emptyset$ to x; moreover, as we write x_t and x_n for denoting a $(t-1)$-dimensional flat X_{t-1} and the whole space $P(n-1,q)$ resp., then we actually apply duality (cf. 3.2) by changing each x_t of x_n into an x_{n-t} of x_n.

Assume $m = \emptyset$. Given any flats $x = x_k$, $y = x_h$, $k \leq h$, consider the flats $x_k \cap x_h = x_i$ and $x_k \vee x_h = x_c$; define

$$d = d(x_k,x_h), \quad \delta_1 = \delta_1(x_k,x_h) = (d-k+h)/2,$$
$$\delta_2 = \delta_2(x_k,x_h) = (d+k-h)/2,$$

so that $d = \delta_1+\delta_2$. Obviously

(3.3) $$d = (c-h)+(c-k) = (h-i)+(k-i),$$

as we can go from x_k to x_h along a geodesic passing through $x_k \vee x_h = x_c$ or along a geodesic passing through $x_k \cap x_h = x_i$. It follows that

$$c = (h+k+d)/2, \quad i = (h+k-d)/2, \quad h-i = \delta_1, \quad c-h = \delta_2.$$

We have to prove that

(3.4) $$\gamma(x_k,x_h) = d!\ [\delta_1]![\delta_2]!/(\delta_1!\delta_2!).$$

If $x_k \subseteq x_h$, then $d = h-k$, $\delta_1 = d$, $\delta_2 = 0$, so that (3.4) reduces to $\gamma(x_k,x_h) = [h-k]!$, which is trivial when $k = h$ and which is also true when $k < h$ because, by duality,

$$\#\{x_{j+1} : x_j \subset x_{j+1} \subseteq x_h\} = \#\{x_{h-j-1} \subset x_{h-j}\} = \#\{x_1 \subset x_{h-j}\}$$
$$= [h-j],$$

$j = k,k+1,\ldots,h-1$.

Now let $x_k \not\subset x_h$ so that $c \geq h+1$. Apply induction on $d \geq 2$. If $d = 2$, then necessarily $h = k$, so that $\delta_1 = \delta_2 = 1$, and (3.4) reduces to $\gamma(x_k,x_h) = 2$, which is true. Assume $d \geq 3$. Obviously

(3.5) $$\gamma(x_k,x_h) = \alpha\,\gamma(x_k,x_{h-1}) + \beta\,\gamma(x_k,x_{h+1}),$$
where

$$\alpha = |\{x_{h-1} : x_i \subseteq x_{h-1} \subset x_h\}|, \quad \beta = |\{x_{h+1} : x_h \subset x_{h+1} \subseteq x_c\}|.$$

By duality, we have

$$\alpha = |\{x_1 \in x_{h-i}\}| = [h-i] = [\delta_1],$$
$$\beta = |\{x_{c-h-1} \subset x_{c-h}\}| = [c-h] = [\delta_2].$$

Moreover

$$d(x_k,x_{h-1}) = d(x_k,x_{h+1}) = d-1,$$
$$\delta_1(x_k,x_{h-1}) = (d-1-k+h-1)/2 = \delta_1-1,$$
$$\delta_2(x_k,x_{h-1}) = (d-1+k-(h-1))/2 = \delta_2,$$
$$\delta_1(x_k,x_{h+1}) = (d-1-k+h+1))/2 = \delta_1,$$
$$\delta_2(x_k,x_{h+1}) = (d-1+k-(h+1))/2 = \delta_2-1.$$

Therefore, by the induction hypothesis, (3.5) becomes:

$$\gamma(x_k,x_h) = [\delta_1]\frac{(d-1)!}{(\delta_1-1)!\,\delta_2!}[\delta_1-1]!\,[\delta_2]!+[\delta_2]\frac{(d-1)!}{\delta_1!\,(\delta_2-1)!}[\delta_1]!\,[\delta_2-1]!$$
$$= \frac{(\delta_1+\delta_2)(d-1)!}{\delta_1!\,\delta_2!}[\delta_1]!\,[\delta_2]! = \frac{d!}{\delta_1!\,\delta_2!}[\delta_1]!\,[\delta_2]!,$$

and (3.4) is proved.

G_3) Given any flats x_k and x_h, consider the flats $x_i = x_k \cap x_h$ and $x_c = x_k \vee x_h$. Then $h+k = i+c$ (Grassmann relation for generalized projective spaces, cf. [8], according to (3.3)). So $h+k-i = c \leq n$, and from (3.3) we get $d(x_k,x_h) = (h-i)+(k-i) \leq n-i$. Therefore $\operatorname{diam}(Q_{n,q}) = \max\{d(x_k,x_h)\} \leq n-\min\{i\} = n$; it follows that $\operatorname{diam}(Q_{n,q}) = n$, because for $x_k = \emptyset$ and $x_h = P$ we have: $d(x_k,x_h) = d(\emptyset,P) = \dim(P)+1 = n$. □

3.5 THEOREM. Let $G = (V,E)$ be a simple graph satisfying conditions G_1), $G_2)_q$, G_3). Then $G \cong Q_{n,q}$.

PROOF. Let $m \in V$ be a vertex as required by condition $G_2)_q$. Consider the partition

$$V = V_0 \cup V_1 \cup \ldots \cup V_N, \quad V_k := \{x \in V : d(m,x) = k\},$$
$$k = 0,\ldots,N.$$

As G is bipartite: $(x,y) \in E \Rightarrow \exists!\ k \in \{0,1,\ldots,N\} : x \in V_k$ and $y \in V_{k\pm1}$. We shall prove that $G \cong Q_{N,q}$ and that $N = n = \operatorname{diam} G$.

Define a *partial order* on V by assuming:

$(x,y \in V)$ $x \leq y$ iff there exists a geodesic $g(m,y)$ containing x.

Consider the map

$$^*: V \to B(V_1) : x \mapsto x^* := \{x_1 \in V_1 : x_1 \leq x\},$$

and define

$$(0 \leq k \leq N) \qquad V_k^* := \{x_k^* : x_k \in V_k\}, \qquad V^* := \text{Im}\ ^*.$$

We shall prove the theorem by the following steps.

(3.5.1). *If* $x,y \in V$ *and* $x \leq y$, *then* $\gamma(x,y) = [d(x,y)]!$: Indeed, $x \leq y$ implies $d(m,y) = d(m,x) + d(x,y)$, then $\delta_1(x,y) = d(x,y)$ and $\delta_2(x,y) = 0$, so that $\gamma(x,y) = [d(x,y)]!$.

(3.5.2). *If* $0 \leq h \leq k$ *and* $x_k \in V_k$, *then* $|\{x_h \in V_h : x_h \leq x_k\}| = C(k,h;q)$: Apply the double counting argument to the set $\{(x_h, g(m,x_k)) : x_h \in g(m,x_k)\}$; by (3.5.1), we get $[k]! = |\{x_h \in V_h : x_h \leq x_k\}|([h]![k-h]!)$, as required.

(3.5.3). $x_k \in V_k \Rightarrow |x_k^*| = [k]$: It follows from (3.5.2) with $h = 1$.

(3.5.4). *Given any two distinct elements* $x_1,y_1 \in V_1$, *there exists exactly one* $x_2 \in V_2$ *such that* $x_1,y_1 \leq x_2$ *(which implies* $x_1,y_1 \in x_2^*$, *and we shall write* $x_1 \vee y_1 = x_2^*$): We have a geodesic (x_1,m,y_1); hence $d(x_1,y_1) = 2$ and $\gamma(x_1,y_1) = 2$. The other geodesic $g(x_1,y_1)$ must be of the type (x_1,x_2,y_1), where $x_2 \in V_2$. In other words there exists exactly one $x_2 \in V_2$ such that $x_1,y_1 \leq x_2$.

(3.5.5). (V_1,V_2^*) *is a linear space of order* q: (3.5.4) gives that (V_1,V_2^*) is a linear space. The order is q, because (3.5.3) implies that for any line z_2^*, $|z_2^*| = [2] = q+1$.

(3.5.6). *The map*

$$^*: V_2 \to V_2^* := \{x_2^* : x_2 \in V_2\} : x_2 \mapsto x_2^*$$

is a bijection: This map is onto by definition and it is injective because if $x_2^* = y_2^*$, then for any distinct points $x_1,y_1 \in x_2^*$ we get $x_1,y_1 \leq x_2$ and $x_1,y_1 \leq y_2$, which implies $x_2 = y_2$, by step (3.5.4).

(3.5.7). *Each* $x_h^* \in V_h^*$ $(0 \le h \le N)$ *is a flat of the linear space* (V_1, V_2^*): By (3.5.2) and (3.5.6), the number of lines $x_2^* \in V_2^*$ contained in x_h^* is more than or equal to the number $C(h,2;q) = |\{x_2 \in V_2 : x_2 \le x_h\}|$. As $|x_h^*| = [h]$, x_h^* contains at most $[h][h-1]/[2] = C(h,2;q)$ lines; then x_h^* contains exactly $C(h,2;q)$ lines, so that for any two distinct points $x_1, y_1 \in V_1$ of x_h^* the line $x_2^* := x_1 \vee y_1$ is contained in x_h^*.

(3.5.8). *Given any* $x_k^* \in V_k^*$, *with* $x_k^* \ne V_1$, *and given any* $y_1 \in V_1 \backslash x_k^*$, $k = 1, \ldots$, *there exists exactly one* $y_{k+1} \in V_{k+1}$ *such that* $y_1 \le y_{k+1}$ *and* $x_k \le y_{k+1}$: First we note that

$$\text{(3.6)} \qquad y_1 \in V_1 \backslash x_k^* \text{ implies } \gamma(y_1, x_k) = (k+1)[k]!,$$

because $d(y_1, x_k) = k+1$, $\delta_1(y_1, x_k) = k$, $\delta_2(y_1, x_k) = 1$. Let ζ denote the number of the required $y_{k+1} \in V_{k+1}$ such that $y_1 \le y_{k+1}$ and $x_k \le y_{k+1}$. Any geodesic $g(y_1, x_k)$ must be either of the type $(y_1, \ldots, x_{k-1}, x_k)$ with $x_{k-1} \le x_k$ and $y_1 \nleq x_{k-1}$, or of the type $(y_1, y_2, \ldots, y_{k+1}, x_k)$ with $y_1 \le y_{k+1}$ and $x_k \le y_{k+1}$. Therefore, by (3.5.2) we have

$$\gamma(y_1, x_k) = C(k,k-1;q) \cdot \gamma(y_1, x_{k-1}) + \zeta \cdot \gamma(y_1, y_{k+1}),$$

so that by (3.5.1) and by (3.6) we get

$$(k+1)[k]! = [k] \cdot k[k-1]! + \zeta \cdot [k]!,$$

which implies $\zeta = 1$.

(3.5.9). *The linear closure of every three non-collinear points of* (V_1, V_2^*) *is a generalized projective plane* x_3^* *of order* q: Given any three non-collinear points $x_1, y_1, z_1 \in V_1$, we have $x_1 \vee y_1 = x_2^*$ and $z_1 \notin x_2^*$. By (3.5.8) $\exists!$ $x_3 \in V_3$ such that $x_1, y_1 \le x_2$ and $z_1, x_2 \le x_3$. Then $x_1, y_1, z_1 \in x_3^*$, which is a flat of (V_1, V_2^*); moreover x_3^* is the linear closure of x_1, y_1, z_1, since x_3^* contains the same number [3] of points belonging to all the lines through z_1 and incident to the line $x_1 \vee y_1 = x_2^*$. Finally, x_3^* is a generalized projective plane: $x_2^* \ne y_2^* \in V_2^*$ and

$x_2^*, y_2^* \subset x_3^*$ imply $x_2^* \cap y_2^* = x_1 \in V_1$, since the number of lines of x_3^* through a point $y_1 \in y_2^*$ is equal to the number [2] of points of x_2^*.

(3.5.10). *(V_1, V_2^*) is a generalized projective space of order q, having V_k^* as the full set of $(k-1)$-flats $(k = 0,1,\ldots)$:* (V_1, V_2^*) is a generalized projective space of order q, by (3.5.5) and (3.5.9). Each $x_k^* \in V_k^*$ is a $(k-1)$-flat, by (3.5.7) and (3.5.3). We must prove that each $(k-1)$-flat is of the type $x_k^* \in V_k^*$. This is trivial for $k = 0,1,2$. Suppose that each $(h-1)$-flat is of the type $x_h^* \in V_h^*$, $h \geq 2$, and let Z_h be an h-flat. Take an $(h-1)$-flat $Z_{h-1} \subset Z_h$ and fix a point $x_1 \in Z_h \setminus Z_{h-1}$. By the induction hypothesis $Z_{h-1} = x_h^*$ for some $x_h^* \in V_h^*$. As $x_1 \notin x_h^*$, $\exists\, x_{h+1} \in V_{h+1}$ such that $x_1, x_h \leq x_{h+1}$ by (3.5.8); so x_{h+1}^* is an h-flat containing x_1 and x_h^*. Therefore $x_{h+1}^* = Z_h$.

(3.5.11). *The dimension of the generalized projective space (V_1, V_2^*) is $N-1$, i.e. it is a generalized projective space $P(N-1,q)$; it follows that there exists exactly one vertex $M = x_N$ at distance N from m: $V_N = \{M\}$ and $M^* = V_1$. Each $x_k \in V_k$ is such that $x_k \leq M$ $(0 \leq k \leq N)$:* The first part immediately follows from (3.5.10) and because N is the maximum value for k. For the second part, we have $d(m,x_k) = k$, and $d(x_k,M) = d(x_k,x_N) = N-k$, by (3.5.10) again; by glueing a geodesic $g(m,x_k)$ with a geodesic $g(x_k,M)$ we get a geodesic $g(m,M)$ containing x_k. So $x_k \leq M$.

(3.5.12). *For any $x_h \in V_h$ and $x_k \in V_k$, we have: $x_h \leq x_k$ iff $x_h^* \subseteq x_k^*$:* If $x_h \leq x_k$, then $x_h^* \subseteq x_k^*$ obviously. Since $|\{x_h \in V_h\colon x_h \leq x_k\}| = C(k,h;q) = |\{x_h^* \in V_h^*\colon x_h^* \subseteq x_k^*\}|$, we have that when x_h describes the set $\{x_h \in V_h\colon x_h \leq x_k\}$, then x_h^* describes the set $\{x_h^* \in V_h^*\colon x_h^* \subseteq x_k^*\}$.

The following step completes the proof of Theorem 3.5.

(3.5.13). *The graph* $G = (V,E)$ *is isomorphic to the graph* $Q_{N,q} = G^* = (V^*,E^*)$ *associated with the generalized projective space* $P(N-1,q) = (V_1,V_2^*)$. *It follows that* $N = \text{diam } G^* = \text{diam } G = n$: The mapping

$$^*: V \equiv V^*: x \mapsto x^* := \{x_1 \in V_1: x_1 \leq x\},$$

is a bijection because each mapping $^*: V_k \to V_k^*$ is a bijection, by (3.5.10). Moreover, by (3.5.12) we have:

$$\begin{aligned}(x^*,y^*) \in E^* &\Leftrightarrow \{(x^* \subset y^*,\ x^* = x_k^*,\ y^* = y_{k+1}^*) \quad \text{or} \\ &\qquad (y^* \subset x^*,\ y^* = y_k^*,\ x^* = x_{k+1}^*)\} \\ &\Leftrightarrow \{(x \leq y,\ x = x_k,\ y = y_{k+1}) \quad \text{or} \\ &\qquad (y \leq x,\ y = y_k,\ x = x_{k+1})\} \\ &\Leftrightarrow (x,y) \in E. \quad \square\end{aligned}$$

REMARK 3.6. We want now to emphasize that for $q = 1$ the characterization of q-hypercubes $Q_{n,q}$ (Thm. 3.4 and Thm. 3.5) gives exactly the Foldes' characterization of hypercubes (Thm. 2.1 and Thm. 2.2): when $q = 1$, we get $Q_{n,q} = Q_n$, $[k] = k$, $[k]! = k!$ and condition $G_2)_q$ turns out to be *equivalent* to condition $G_2)$.

COROLLARY 3.7. A graph $G = (V,E)$ is isomorphic to the graph of the lattice of the subspaces of a vector space of a given finite dimension $n \geq 4$ over a finite field iff G satisfies conditions $G_1)$, $G_2)_q$ and $G_3)$, for some $q \geq 2$.

PROOF. A generalized projective space of dimension $d \geq 3$ and of order $q \geq 2$ is necessarily a desarguesian projective space and so it is the projective space PG(d,q) over the Galois field GF(q). The lattice of the subspaces of PG(d,q) is isomorphic to the lattice of the subspaces of the vector space of dimension d+1 over GF(q). So the result immediately follows from Thm. 3.4 and Thm. 3.5. □

3.8 GEOMETRIC SPACES. A *geometric space* is defined as a pair $S = (P,B)$, where P is a finite set of elements, called *points* of S, and B is a set of distinguished subsets of P, called *flats* of

S, with the property that $\emptyset$, P and every singleton of P belong to B. A flat x is called a *maximal subflat* of a given flat y if

$$x \subsetneq y \text{ and } (\forall z \in B)\ x \subseteq z \subseteq y \text{ implies either } z = x \text{ or } z = y.$$

The *graph associated to the geometric space* $S = (P,B)$ is defined as the graph $G(S) = (V,E)$ where $V = B$, and $(x,y) \in E$ iff either x is a maximal subflat of y or y is a maximal subflat of x. As $P \in B$, the graph G(S) associated to a geometric space S is necessarily connected. From Thm. 3.4 and Thm. 3.5 we get the following corollary.

3.9 COROLLARY. Let $S = (P,B)$ be a geometric space, and let $G(S) = (V,E)$ be the associated graph. Then S is a generalized projective space having P as the set of points and B as the set of flats iff G is bipartite and there exists an integer $q \geq 1$ with the property that for any two vertices $x,y \in V$ the number of geodesics $g(x,y)$ is given by (3.1) where $m = \emptyset$. □

4. GEODESICS IN A PRODUCT GRAPH

Let $G_i = (V_i,E_i)$ be a connected graph with distance function d_i, $i = 1,2,\ldots,n$. Let us consider the cartesian product $V = V_1 \times V_2 \times \ldots \times V_n$ and define:

$$(\forall\ x = (x_i),\ y = (y_i) \in V) \qquad d_i(x,y) := d_i(x_i,y_i).$$

The (cartesian) *product graph* $G = \Pi G_i = (V,E)$ is defined by:

$$(x,y) \in E \quad \text{iff} \quad \textstyle\sum_i d_i(x,y) = 1.$$

The distance in the graph G is then given by:

$$(\forall\ x,y \in V) \qquad d(x,y) = \textstyle\sum_i d_i(x,y).$$

For any $x,y \in V$, and for any $i = 1,2,\ldots,n$, we shall define:

$\Gamma_G(x,y)$:= the set of geodesics $g(x,y)$ of G between x and y,

$\Gamma_i(x,y) := \Gamma_i(x_i,y_i)$:= the set of geodesics $g_i(x,y) = g_i(x_i,y_i)$ of G_i between x_i and y_i,

$\gamma_G(x,y) := |\Gamma_G(x,y)|, \quad \gamma_i(x,y) := \gamma_i(x_i,y_i) := |\Gamma_i(x,y)|.$

4.1 THEOREM. Let $G = \Pi G_i = (V,E)$ be a product graph $(i = 1,\ldots,n)$. Then

$$(\forall\, x,y \in V)$$

$$\gamma_G(x,y) = (\Pi_i \gamma_i(x,y)) \cdot \left[\Sigma_i\, d_i(x,y)\right]! / \Pi_i (d_i(x,y)!).$$

PROOF. A geodesic $g(x,y)$ of G is given iff we give:

(a) an element $(g_1,g_2,\ldots,g_n) \in \Pi_i \Gamma_i(x_i,y_i)$; i.e. if we give a geodesic $g_i = g_i(x_i,y_i)$ for each i-component;

(b) an ordered d-arrangement of $d = \Sigma_i\, d_i(x,y)$ objects, of which exactly $d_1 = d_1(x,y)$ are like g_1, exactly $d_2 = d_2(x,y)$ are like $g_2,\ldots,$ and exactly $d_n = d_n(x,y)$ are like g_n; i.e. if we fix, for each of the d ordered steps from x to y along the required geodesic $g(x,y)$, the j-th component to be moved of one step along the corresponding geodesic $g_j(x_j,y_j)$, $j = 1,\ldots,n$.

From (a) and (b) it follows that there is a bijection between the set $\Gamma_G(x,y)$ and the set $A(x,y) \times B(x,y)$, where $A(x,y) := \Pi_i \Gamma_i(x,y)$ and $B(x,y) :=$ {ordered d-arrangements of d objects, of which exactly d_1 are like $g_1,\ldots,$ and exactly d_n are alike g_n}. This proves the theorem. □

Let $G = (V,E)$ be any graph and define

$$D(G) := \{d(x,y)\colon x,y \in V\} = \{0,1,2,\ldots,\text{diam } G\}.$$

The graph G will be called a *function geodetic graph* if the number of geodesics of G between any two vertices depends only on their distance, i.e. if there exists a map $F\colon D(G) \to \mathbf{N}$ such that

$$(\forall\, x,y \in V)\ \gamma_G(x,y) = F(d(x,y));$$

in this case we shall also say that G is an *F-geodetic graph* (cf. [3], [4], [7]).

In particular, when G is F-geodetic with $F(h) = 1$ for every $h \in D(G)$, i.e. when $\gamma_G(x,y) = 1$ for every $x,y \in V$, then G is called a *geodetic graph* (cf. [9]); when G is F-geodetic with $F(h) = h!$ for every $h \in D(G)$, i.e. when $\gamma_G(x,y) = d(x,y)!$ for every $x,y \in V$, then G will be called a *factorial geodetic graph* (an *interval regular graph* in the terminology of [6]). As $h! = 1$ implies $h \in \{0,1\}$, it is clear that:

4.2 PROPOSITION. The intersection between the class of geodetic graphs and the class of factorial geodetic graphs is the class of complete graphs. □

It immediately follows, from Thm. 4.1, that any product of complete graphs, i.e. any *Hamming graph*, in particular any *cube* $Q_n = (K_2)^n$, is a factorial geodetic graph.

If G is a *product graph*, $G = \Pi G_i = (V,E)$, $i = 1,2,\ldots,n$, and if $x,y \in V$, we shall define the *distance vector* δ: $V \times V \to \mathbf{N}^n$ by assuming

$$(\forall x,y \in V)\ \delta(x,y) := (d_1(x,y),\ldots,d_n(x,y)),$$

and we shall say that G is a *distance vector geodetic graph*, if the number of geodesics of G between any two vertices depends only on their distance vector, i.e. if there exists a map

$$\Phi\colon D(G_1) \times D(G_2) \times\ldots\times D(G_n) \to \mathbf{N}.$$

such that: $(\forall x,y \in V)\ \gamma_G(x,y) = \Phi(\delta(x,y))$; in this case we shall also say that G is a (Φ,δ)-*geodetic graph*.

A graph G will be called a *multinomial geodetic graph* if it is a product graph and it is a (Φ,δ)-geodetic graph, with $\Phi(h_1,\ldots,h_n) = \left[\Sigma_i h_i\right]!/\Pi_i(h_i!)$.

4.3 THEOREM. A graph G is a multinomial geodetic graph iff it is the product of geodetic graphs.

PROOF. Let $G = \Pi G_i = (V,E)$, $i = 1,2,\ldots,n$. From Thm. 4.1 we have that:

$$G \text{ is multinomial geodetic} \Leftrightarrow (\forall x,y \in V)\ \Pi_i\gamma_i(x,y) = 1$$
$$\Leftrightarrow (\forall i)(\forall x_i,y_i \in V_i)\ \gamma_i(x_i,y_i) = 1 \Leftrightarrow (\forall i)\ G_i \text{ is geodetic.}$$ □

4.4 THEOREM. A product graph is function geodetic iff it is factorial geodetic iff each factor is factorial geodetic.

PROOF. Let $G = \Pi G_i = (V,E)$ be a product graph $(i = 1,\ldots,n)$. If $(\forall i)$ G_i is factorial geodetic, i.e. if $(\forall i)(\forall x_i,y_i \in V_i)$ $\gamma_i(x_i,y_i) = d_i(x_i,y_i)!$, then Thm. 4.1 gives that $(\forall x,y \in V)$ $\gamma_G(x,y) = \left[\Sigma_i d_i(x,y)\right]! = d(x,y)!$, i.e. G is factorial geodetic and so it is function geodetic.

Conversely, suppose that G is F-geodetic. First of all we observe that each G_i is F_i-geodetic where F_i is the restriction of F to $D(G_i)$. Indeed, if $x_i, y_i \in V_i$, extend these vertices of V_i to vertices $x,y \in V$ by assuming all the other components x_j of x and y_j of y both to be equal to a given $z_j \in V_j$: $x_j = y_j = z_j$, $j = 1,\ldots,n$, $j \neq i$. As $d(x,y) = d_i(x,y)$ and $d_j(x,y) = 0$ for $j \neq i$, we get from Thm. 4.1 that $\gamma_G(x,y) = \gamma_i(x,y)$, which also follows because there is a natural bijection between $\Gamma_i(x,y)$ and $\Gamma_G(x,y)$; so we have $\gamma_i(x,y) = \gamma_G(x,y) = F(d(x,y)) = F(d_i(x,y))$. This proves that G_i is F_i-geodetic, where F_i is the restriction of F to $D(G_i)$. From Thm. 4.1, we get that F must satisfy the condition:

$$(\forall i)(\forall\ h_i \in D(G_i))\ F\left[\Sigma_i\ h_i\right] = (\Pi_i F(h_i))\cdot\left[\Sigma_i\ h_i\right]!/\Pi_i(h_i!).$$

This equality implies $F(h) = h!$ for all $h \in D(G)$, as immediately follows by an induction argument on h. This completes the proof. □

4.5 THEOREM. A graph G is a Hamming graph iff it is multinomial geodetic and factorial geodetic.

PROOF. By Thm. 4.3 and Thm. 4.4, we have that G is multinomial geodetic and factorial geodetic iff $G = \Pi G_i$ where each factor G_i is geodetic and also factorial geodetic; by Thm. 4.2 this means that each factor G_i is a complete graph, i.e., by definition, that ΠG_i is a Hamming graph. □

4.6 THEOREM. Let $G = \Pi G_i = (V,E)$ be a product graph, and let $(h_1,\ldots,h_n) \in \Pi_i D(G_i)$, $(i = 1,\ldots,n)$. Then:

(a) If G is (Φ,δ)-geodetic, then each G_i is F_i-geodetic with

$$F_i(h_i) = \Phi(0,\ldots,0,h_i,0,\ldots,0).$$

(b) If every G_i is F_i-geodetic, then G is (Φ,δ)-geodetic with

$$\Phi(h_1,\ldots,h_n) = (\Pi_i F_i(h_i))\cdot\left[\Sigma_i\ h_i\right]!/\Pi_i(h_i!).$$

PROOF. (a) For a fixed i, let $x_i, y_i \in V_i$ such that $d_i(x,y) =$

h_i and let $x,y \in V$ be vertices of G defined as in the second part of the proof of Thm. 4.4; the same argument used in that case gives $\delta(x,y) = (0,\ldots,h_i,\ldots,0)$, and

$$\gamma_i(x,y) = \gamma_G(x,y) = \Phi(\delta(x,y)) = \Phi(0,\ldots,h_i,\ldots,0).$$

This proves that G_i is F_i-geodetic with $F_i(h_i) = \Phi(0,\ldots,h_i,\ldots,0)$.

(b) Given any $(h_1,\ldots,h_n) \in \Pi_i D(G_i)$, fix any x and y in V such that $d_i(x,y) = h_i$, i.e. such that $\delta(x,y) = (h_1,\ldots,h_n)$. As G_i is F_i-geodetic, we have $\gamma_i(x_i,y_i) = F_i(d_i(x,y)) = F_i(h_i)$, so that $\gamma_G(x,y) = (\Pi_i\gamma_i(x_i,y_i))\cdot\left[\sum_i d_i(x,y)\right]!/\Pi_i(d_i(x,y)!) = (\Pi_i F_i(h_i))\cdot\left[\sum_i h_i\right]!/\Pi_i(h_i!)$. This proves that G is (Φ,δ)-geodetic with $\Phi(h_1,\ldots,h_n) = (\Pi_i F_i(h_i))\cdot\left[\sum_i h_i\right]!/\Pi_i(h_i!)$.

□

REFERENCES

[1] P.V. Ceccherini, A q-analogue of the characterization of hypercubes as graphs, J. Geometry, 22(1984), 57-74.

[2] P.V. Ceccherini and A. Sappa, A new characterization of hypercubes, Ann. Discrete Math. 30(1986), 137-142.

[3] P.V. Ceccherini and A. Sappa, F-binomial coefficients and related combinatorial topics: perfect matroid designs, parti ordered sets of full binomial type and F-geodetic graphs, An Discrete Math. 30(1986), 143-158.

[4] R.J. Cook, D.G. Pryce, Uniformly geodetic graphs, Ars Combinatoria 16-A(1983), 55-59.

[5] S. Foldes, A characterization of hypercubes, Discrete Math. 17(1977), 155-159.

[6] H.M. Mulder, Interval regular graphs, Discrete Math. 41(1982 253-269.

[7] R. Scapellato, Geodetic graphs of diameter two and some rela structures, J. Combinatorial Theory 41B(1986), 218-229.

[8] B. Segre, Lectures on modern geometry. With an Appendix by L Lombardo-Radice. Cremonese, Roma, 1961.

[9] J.C. Stemple, Geodetic graphs of diameter two, J. Combinator Theory 17B(1974), 266-280.

Existence of (v,8,1)-BIBD

DU Beiliang Department of Mathematics, Suzhou University, Suzhou, China

ZHU Lie, Department of Mathematics, Suzhou University, Suzhou, China

1. INTRODUCTION

DEFINITION 1.1. Let v, k and λ be positive integers. A *balanced incomplete block design* with parameters v, k, λ (briefly (v,k,λ)-BIBD) is a pair $(X,\mathcal{B})$ where X is a v-set (of *points*) and $\mathcal{B}$ is a collection of k-subsets of X (called *blocks*) such that every pair of points of X appears in exactly λ blocks of $\mathcal{B}$.

Some simple computation shows, for any given integers v, k and λ, a (v,k,λ)-BIBD exists only if

$$\begin{aligned} \lambda(v-1) &\equiv 0 \pmod{k-1} \\ \lambda v(v-1) &\equiv 0 \pmod{k(k-1)}. \end{aligned} \tag{1.1}$$

For $k \leq 5$ and $k = 6$, $\lambda > 1$, Hanani [6] has shown that a BIBD exists for all v, k, λ satisfying the condition (1.1) with the exception of $(v,k,\lambda) = (15,5,2)$ and $(21,6,2)$. For $k = 6$, and $\lambda = 1$, Mills [10] has proved that the necessary condition (1.1) is also sufficient for all $v > 11151$. The cases when $v = 16$, 21 or 36 are impossible and there are 165 values below 11151 for which the existence of a $(v,6,1)$-BIBD is undecided. Recently, 64 values of v have been decided affirmatively in [11] and [17]. Mills has since decided another six values (see [12]). Therefore, there are 95 values of v left undecided in which 5901 is the largest. For $k \geq 7$, not much work has been done. Hanani [6] has proved that the necessary existence condition (1.1) for $(v,7,\lambda)$-BIBD is sufficient provided that $\lambda \equiv 0, 6, 7, 12, 18, 24, 30, 35$, or 36 (mod 42) or $\lambda > 30$ with $\gcd(\lambda,6) = 1$. In this paper we shall investigate the case $k = 8$ and $\lambda = 1$.

When $k = 8$ and $\lambda = 1$, the necessary condition (1.1) is equivalent to

$$v \equiv 1 \text{ or } 8 \pmod{56}. \tag{1.2}$$

We shall show that a $(v,8,1)$-BIBD exists for all v satisfying the condition (1.2) with at most 223 possible exceptions, of which 21897 is the largest. That is

THEOREM 1.2. A $(v,8,1)$-BIBD exists whenever $v \equiv 1$ or 8 (mod 56) with 223 possible exceptions of $v = 7r+1$, where r is shown in Table 1, and the notation "a ~ b" denotes the set $\{r: a \leq r \leq b, r \equiv 0 \text{ or } 1 \pmod 8\}$.

TABLE 1

16	24	25	32	33	40	41	48	56	88
89	96	97	104	105	112	113	"160 ~	272"	280
281	288	296	297	304	305	312	313	320	321
328	329	336	337	360	368	369	376	377	384
416	424	425	432	440	472	480	481	488	489
496	536	544	545	552	608	744	745	752	753
760	761	768	776	808	809	816	817	824	825
832	840	849	856	857	864	865	872	873	880
881	888	889	896	920	928	929	936	937	944
952	984	992	993	1000	1001	1056	"1528 ~	1857"	1880
1881	1888	1889	1896	1897	1904	1905	1952	1960	1968
3064	3065	3072	3128						

This result is established, for the most part, by means of a result on pairwise balanced designs (PBDs) and which is of interest in its own right.

DEFINITION 1.3. Let K be a set of positive integers. A *pairwise balanced design* (PBD) of index unity B(K,1; r) is a pair $(X,\mathcal{B})$ where X is an r-set (of *points*) and $\mathcal{B}$ is a collection of subsets of X (called *blocks*) with sizes from K such that every unordered pair of distinct points of X is contained in exactly one block of $\mathcal{B}$. The number $|X| = r$ is called the *order* of the PBD.

If Q denotes the set {8,9,17,49}, then it is shown that a PBD B(Q,1; r) exists for all integers $r \equiv 0$ or 1 (mod 8), with at most 226 possible exceptions of r = 385, 448, 3696 and the values listed in Table 1. That is

THEOREM 1.4. Let Q denote the set {8,9,17,49}. Then a PBD B(Q,1; r) exists for all positive integers $r \equiv 0$ or 1 (mod 8), with the possible exceptions of r = 385, 448, 3696 and the 223 values listed in Table 1.

2. PRELIMINARIES

In this section, we shall define some terminology and state some fundamental results which will be used later. For more detailed information on PBDs and related designs, the interested reader may refer to [1, 6, 14, 16].

DEFINITION 2.1. Let K and M be sets of positive integers. A *group divisible design* (GDD) GD(K,1,M;v) is a triple $(X,\mathcal{G},\mathcal{B})$, where

(i) X is a v-set (of *points*),

(ii) $\mathcal{G}$ is a collection of non-empty subsets of X (called *groups*) with sizes in M and which partition X,

(iii) $\mathcal{B}$ is a collection of subsets of X (called *blocks*), each with size at least two in K,

(iv) no block meets a group in more than one point, and

(v) each pairset {x,y} of points not contained in a group

is contained in exactly one block.
The *group-type* (or *type*) of a GDD($X,\mathcal{G},\mathcal{B}$) is the multiset $\{|G|: G \in \mathcal{G}\}$ and we shall use the "exponential" notation for its description: a group-type $1^i 2^j 3^k \cdots$ denotes i occurrences of groups of size 1, j occurrences of groups of size 2, and so on. A weighting of a GDD ($X,\mathcal{G},\mathcal{B}$) is any mapping $w: X \to Z^+ \cup \{0\}$.

DEFINITION 2.2. A *transversal design* (TD) T(k,1; m) is a GDD with km points, k groups of size m and m^2 blocks of size k where each block meets every group in precisely one point, that is, each block is a transversal of the collection of groups.

DEFINITION 2.3. Let ($X,\mathcal{B}$) be a PBD B(K,1;v). A *parallel class* in ($X,\mathcal{B}$) is a collection of disjoint blocks of $\mathcal{B}$, the union of which equals X. ($X,\mathcal{B}$) is called *resolvable* if the blocks of $\mathcal{B}$ can be partitioned into parallel classes. A GDD GD(K,1,M;v) is resolvable if its associated PBD B($K \cup M$,1;v) is resolvable with M as a parallel class of the resolution.

It is fairly well-known that the existence of a resolvable TD T(k,1;m) (briefly RT(k,1;m)) is equivalent to the existence of a T(k+1,1;m) or equivalently k-1 mutually orthogonal Latin squares (MOLS) of order m. Moreover, the following two results can be found in [8].

THEOREM 2.4. For every prime power q, there exists a T(q+1,1; q).

THEOREM 2.5. Let $m = p_1^{k_1} p_2^{k_2} \cdots p_r^{k_r}$ be the factorization of m into powers of distinct primes p_i, then a T(k,1; m) exists, where $k \le 1+\min\{p_i^{k_i}\}$.

We shall denote by N(m) the maximum number of MOLS of order m. From current results we have on the existence of sets of t MOLS (see, for example, [3, 4]), in particular, we can state the following useful result.

THEOREM 2.6. $N(m) \ge 7$ for any $m \ge 781$.

We shall denote by $B(K)$ the set of all integers v for which there exists a PBD $B(K,1;v)$. For convenience, we write $B(k_1,k_2,\cdots,k_r)$ for $B(\{k_1,k_2,\cdots,k_r\})$. A set K is said to be *PBD-closed* if $B(K) = K$.

We shall briefly write $B(k,1;v)$ for $B(\{k\},1;v)$ and similarly $GD(k,1,m;v)$ for $GD(\{k\},1,\{m\};v)$. We also observe that a PBD $B(k,1;v)$ is essentially a $(v,k,1)$-BIBD. If $k \notin K$, then $B(K \cup \{k^*\},1;v)$ denotes a PBD $B(K \cup \{k\},1;v)$ which contains a unique block of size k and if $k \in K$, then $B(K \cup \{k^*\},1;v)$ is a PBD $B(K,1;v)$ containing at least one block of size k. We shall sometimes refer to a PBD $B(K,1;v)$ as a $(v,K,1)$-PBD, and a GDD $(x,\mathcal{G},\mathcal{B})$ will be referred to as a K-GDD if $|B| \in K$ for every block in $\mathcal{B}$. We shall also adapt the following notations:

$$B(k) = \{v\colon \text{a } (v,k,1)\text{-BIBD exists}\},$$
$$RB(k) = \{v\colon \text{a resolvable } (v,k,1)\text{-BIBD exists}\},$$
$$R_k = \{r\colon (k-1)r+1 \in B(k)\},$$
$$R_k^* = \{r\colon (k-1)r+1 \in RB(k)\}.$$

It is known [16] that for any $k > 2$, the sets R_k and R_k^* are PBD-closed. Moreover, if k is a prime power, then it is known that $k+1 \in R_k^*$ and so we have $B(k+1) \subset R_k^*$. We also point out that a $B(q+1,1;q^2+q+1)$ is a projective plane of order q, denoted by $PG(2,q)$, and a $B(q,1;q^2)$ is an affine plane of order q, denoted by $AG(2,q)$. An *oval* of an affine or projective plane of order q is a set of $q+1$ points, no three of which are collinear. The following useful result is proved in [1, VIII, 9.12 Theorem], and similar constructions can be found in [3, 14].

THEOREM 2.7. Let q be a prime power. Then there exist both an affine and a projective plane of order q with an oval. Hence for $0 \le t \le q+1$, we have

(i) $q^2-t \in B(q, q-1, q-2)$,

(ii) $q^2+q+1-t \in B(q+1, q, q-1)$.

The following result is due to Bose [2].

THEOREM 2.8. Let q be a prime power. Then $q^3+1 \in RB(q+1)$.

The following result is due to Hanani, Ray-Chaudhuri and Wilson [7].

THEOREM 2.9. Suppose q is a prime power and n is any positive integer. Then $q^n \in RB(q)$.

The following construction is due to Seiden [13].

THEOREM 2.10. For any positive integer n, we have $2^{2n-1} - 2^{n-1} \in RB(2^{n-1})$.

We shall also make use of the following result, which is contained in [9].

THEOREM 2.11. Let q be a prime power. Then $q^3+q^2+q+1 \in RB(q+1)$.

LEMMA 2.12. $Q \subset R_8$, where $Q = \{8,9,17,49\}$.

PROOF. From a projective plane of order 7 and affine plane of order 8 we have $8,9 \in R_8$. Applying Theorem 2.10 with $n = 4$ gives $120 \in RB(8)$ and $17 \in R_8$. Applying Theorem 2.8 with $q = 7$ gives $344 \in RB(8)$ and $49 \in R_8$.

Since R_8 is PBD-closed, we readily obtain

LEMMA 2.13. $B(Q) \subset R_8$.

For some of our recursive constructions of PBDs and GDDs, we shall make use of Wilson's "Fundamental Construction" (see [16]). A brief description is presented below.

CONSTRUCTION 2.14 (Fundamental Construction). Suppose that $(X,\mathcal{G},\mathcal{B})$ is a "master" GDD and let $w: X \to Z^+ \cup \{0\}$ be a weighting of the GDD. For every $x \in X$, let s_x be $w(x)$ "copies" of x. Suppose that for each block $B \in \mathcal{B}$, a GDD $(\cup_{x\in B}s_x, \{s_x: x \in B\}, \mathcal{A}_B)$ is given. Let $X^* = \cup_{x\in X}s_x$, $\mathcal{G}^* = \{\cup_{x\in G}s_x: G \in \mathcal{G}\}$, $\mathcal{B}^* = \cup_{B\in\mathcal{B}}\mathcal{A}_B$. Then $(X^*,\mathcal{G}^*,\mathcal{B}^*)$ is a GDD.

As already mentioned, our main result will be established on the basis of our investigation of the set B(Q). It will be convenient for us to proceed in stages. Accordingly, we define the following sets:

$$D_0 = \{d:\ 8d \in B(Q)\},$$
$$D_1 = \{d:\ 8d+1 \in B(Q)\},$$
$$D = D_0 \cap D_1.$$

Since D is not PBD-closed, we shall consider the following set $U \subset D$ which is PBD-closed and defined by:

$$U = \{u:\ \text{there exists a } GD(B(Q),1,8;8u)\}.$$

In what follows, we shall first investigate the sets U, D_1 and D_0, and then apply Lemma 2.13 to obtain our main result in Theorem 1.2.

3. DETERMINATION OF U

The following lemma is fairly obvious.

LEMMA 3.1. If $u \in B(Q)$ and $N(u) \geq 7$, then $u \in U$. In particular, $Q \subset U$.

LEMMA 3.2. If $v \in B(9)$, then $(v-e)/8 \in U$ where e = 1 or 9.

PROOF. In a (v,9,1)-BIBD, we delete e points from a particular block to obtain a GD({8,9},1,8; v-e). The conclusion follows.

LEMMA 3.3. $\{81,153,513,1873,2017\} \subset B(9)$.

PROOF. Apply Theorem 2.9 with q = 9, n = 2 to obtain $81 \in B(9)$. Apply Theorem 2.8 with q = 8 to obtain $513 \in B(9)$. For $153 \in B(9)$ see Appendix I in [5] and for $1873,2017 \in B(9)$ see [15].

As an immediate consequence of Lemmas 3.2 and 3.3, we obtain

COROLLARY 3.4. $\{10,18,19,63,64,233,234,251,252\} \subset U$.

LEMMA 3.5. If $v \in RB(8)$, then $v/8+u \in U$, where u = 0 or $u \in U$

and $8u < (v-1)/7$.

PROOF. For $u = 0$ we have $v/8 \in U$ since $v \in RB(8)$. Now we assume $u > 0$ including $u = 1$ which is contained in U. We adjoin $8u$ infinite points to a resolvable $(v,8,1)$-BIBD, where one infinite point is adjoined to each of $8u$ parallel classes of blocks. In the resulting design, we then take as groups the blocks of one the remaining parallel classes together with the block at infinity of size $8u$ to obtain a $\{8,9\}$-GDD of group-type $8^{v/8}(8u)^1$. Since $u \in U$, we can then break up the group of size $8u$ to form a $GD(B(Q),1,8;\ 8u)$ and obtain $v/8+u \in U$.

LEMMA 3.6. $\{120,344,400,512,568,960,1016,1072,1912,2024\} \subset RB(8)$.

PROOF. $120,344 \in RB(8)$ comes from the proof of Lemma 2.12. Applying Theorem 2.11 with $q = 7$ we obtain $400 \in RB(8)$, and applying Theorem 2.9 with $q = 8$, $n = 3$ we get $512 \in RB(8)$. Since $64,120 \in RB(8)$ implies $9,17 \in R_8^*$, and $81,153 \in B(9)$ from Lemma 3.3, so $81,153 \in R_8^*$ and $568,1072 \in RB(8)$. We adjoin 17 infinite points to a resolvable $(120,8,1)$-BIBD, where one infinite point is adjoined to each of 17 parallel classes of blocks, to obtains $(137,\{9,17\},1)$-PBD and $137 \in R_8^*$, so $960 \in RB(8)$. We adjoin one infinite point to a $T(9,1;16)$, to obtain $(145,\{9,17\},1)$-PBD and $145 \in R_8^*$, so $1016 \in RB(8)$. From a projective plane and an affine plane we have $273,289 \in B(17) \subset R_8^*$ and then $1912,2024 \in RB(8)$.

As an immediate consequence of Lemmas 3.5 and 3.6, we obtain

COROLLARY 3.7. $\{15,16,43,44,50,51,65,71,72,73,79,80,120, 121,127,134,239,240,247,248,249,253,254,255\} \subset U$.

LEMMA 3.8. If $N(t) \geq 7$ and $t+q \in B(Q)$ where $q = 0$ or 1, then $t+u \in U$ provided $u \in U$ and $q \leq 8u < t$.

PROOF. We take an $RT(8,1;\ t)$ and adjoin $8u-q$ infinite points to

$8u-q$ parallel classes of blocks and $q = 0$ or 1 infinite point to the groups so as to form a $\{8,9,t+q\}$-GDD of group-type $8^t(8u)^1$. Using the fact that $t+q \in B(Q)$ and $u \in U$, we can then break up the size $t+q$ blocks and the group of size $8u$ to form a $B(Q)$-GDD of group-type $8^t 8^u$ and obtain $t+u \in U$.

LEMMA 3.9. $\{57,65,73,121\} \subset B(Q)$.

PROOF. From the existence of a projective plane we know that $57 \in B(Q)$. Since $\{8,9,15\} \subset U$, we know from the corresponding GDDs that $\{65,73,121\} \subset B(Q)$.

As an immediate consequence of Lemmas 3.8 and 3.9, where the condition $N(t) \geq 7$ is met from [3], we obtain

COROLLARY 3.10. $\{57,58,66,74,122\} \subset U$.

The following lemma is an immediate consequence of Theorem 2.7.

LEMMA 3.11. If $81 \leq u \leq 91$, then $u \in U$.

PROOF. We apply Theorem 2.7 with $q = 9$ in (ii) to obtain $u \in B(8,9,10)$ for $81 \leq u \leq 91$. Since $\{8,9,10\} \subset U$ and U is PBD-closed, the conclusion follows.

LEMMA 3.12. $\{92,98,99,100,141\} \subset U$.

PROOF. In a $T(10,1;11)$, we consider two disjoint blocks and delete all but one of the points in each block in such a way that the two remaining points did not lie in the same group. Then we obtain $92 \in B(8,9,10)$, and $92 \in U$. If we delete one block and some points in a group of a $T(10,1;11)$ in such a way that the small group is of size 8, 9, or 10, then we get $\{98,99,100\} \subset U$. In a $T(15,1;16)$, we select two disjoint blocks: delete all the points in one block, and then in the last six groups of the resulting design we keep only the points in the second selected block and delete all other points, to obtain $141 \in B(8,9,10,15)$ and $141 \in U$.

LEMMA 3.13. If $N(t) \geq 7$ and $t+8 \in B(Q \cup 8^*\})$, then $t+1 \in U$.

PROOF. We take an $RT(8,1;t)$ and adjoin 8 infinite points to it by forming a $(t+8, Q \cup \{8^*\},1)$-PBD on each group in such a way that the 8 infinite points become a common block in these PBDs. Taking a parallel class of blocks in the $RT(8,1;t)$ together with the block at infinity of size 8 as groups we obtain a $B(Q)$-GDD of group-type 8^{t+1}. That is, $t+1 \in U$.

COROLLARY 3.14. $\{113,114,385\} \subset U$.

PROOF. Take $t = 112,113,384$ in Lemma 3.13. From [3] we have $N(t) \geq 7$. The conditions $120 \in B(8)$ and $121 \in B(8,9)$ come from $120 \in RB(8)$, and $392 \in B(Q \cup \{8^*\})$ comes from $49 \in U$.

LEMMA 3.15. Suppose $t \in U$ and t is a prime power greater than 9. Suppose $8+b \in U$ and $8+b \leq t$. Then $8t+a+b \in U$ if $a \in U$ and $0 \leq a < t$; or $8t+a+b-1 \in U$ if $a \in U$ and $0 < a \leq t$.

PROOF. In a $T(t+1,1;t)$, we take a particular block B of size $t+1$. Keeping a point in the ninth group of the TD, we delete the other points including the point of intersection with B. Keeping the points in the first eight groups of the TD and b additional points in B, we delete all the other points in the last $t-8$ groups so as to obtain an $(8t+a+b,U,1)$-PBD, that is, $8t+a+b \in U$. If we leave the point of intersection of B and the ninth group undeleted, then we get $8t+a+b-1 \in U$.

LEMMA 3.16. Suppose $N(t) \geq 8$ and $q \in \{0,1\}, u,v \leq t$. Then

$$\{t+q,u+q,v+q\} \subset U \text{ implies } 8t+u+v+q \in U.$$

PROOF. In a $T(10,1;t)$, we delete $t-u$ points from one group and $t-v$ points from another group to obtain an $\{8,9,10\}$-GDD of group-type $t^8u^1v^1$. We then adjoin q infinite points to the groups of this GDD in order to obtain the desired result.

LEMMA 3.17. If $128 \leq n \leq 149$ and $n \neq 132,133$, or if $171 \leq n \leq 179$, then $n \in U$.

PROOF. We apply Lemma 3.16 with t = 16,19 to get $8t+u+v+q \in U$, where the required parameters are shown in Table 2. In addition, we can take $(a,b) \in \{(1,2),(0,7),(10,2),(8,7),(15,7)\}$ in Lemma 3.15 to obtain $8t+m \in U$ where m = 3,7,12,14,21,22. In particular, we have $8\cdot16+m \in U$ for $0 \le m \le 21$, $m \ne 4,5,6,13$, and $8\cdot19+m \in U$ for $19 \le m \le 27$. All the values of n have been covered except for n = 134,141, which are taken care of by Corollary 3.7 and Lemma 3.12.

TABLE 2

u+v+q	u	v	q	u+v+q	u	v	q
0	0	0	0	29	19	10	0
1	1	0	0	30	15	15	0
2	1	1	0	31	16	15	0
3	—	—	—	32	16	16	0
4	—	—	—	33	17	16	0
5	—	—	—	34	17	17	0
6	—	—	—	35	18	17	0
7	—	—	—	36	18	18	0
8	8	0	0	37	19	18	0
9	8	1	0	38	19	19	0
10	9	1	0	39	—	—	—
11	10	1	0	40	—	—	—
12	—	—	—	41	—	—	—
13	—	—	—	42	—	—	—
14	—	—	—	43	43	0	0
15	15	0	0	44	43	1	1
16	8	8	0	45	44	1	0
17	9	8	0	46	—	—	—
18	9	9	0	47	—	—	—
19	10	9	0	48	—	—	—
20	10	10	0	49	49	0	0
21	—	—	—	50	42	7	1
22	14	7	1	51	43	8	0
23	15	8	0	52	43	9	0
24	15	9	0	53	43	10	0
25	15	10	0	54	44	10	0
26	16	10	0	55	—	—	—
27	17	10	0	56	—	—	—
28	18	10	0				

DEFINITION 3.18. An *oval in a group divisible design* is a set of points which intersects each group in at most one point and each block in at most two points.

As an immediate consequence of Theorem 2.7, we obtain

LEMMA 3.19. Let q be a prime power. Then there exists an oval of size m in a $T(m,1;q)$, where $m \leq q$.

LEMMA 3.20. If $150 \leq n \leq 170$ or $180 \leq n \leq 190$, then $n \in U$.

PROOF. Apply Lemma 3.19 with $q = 16,17,19$ and $m = 10$ to obtain an oval of size 10 in $T(10,1;q)$. By deleting k points from the oval, we get $10q-k \in B(8,9,10,q-1,q) \subset U$, where $0 \leq k \leq 10$.

LEMMA 3.21. Suppose $N(t) \geq 15$ and $q \in \{0,1\}$, $u,v \leq t$. Then

$$\{t+q,u+q,v+q\} \subset U \text{ implies } 15t+u+v+q \in U.$$

PROOF. In a $T(17,1;t)$ we delete t-u points from one group and t-v points from another group to obtain a {15,16,17}-GDD of group-type $t^{15}u^1v^1$. We then adjoin q infinite points to the groups of this GDD in order to obtain the desired result.

COROLLARY 3.22. $\{241,242,250\} \subset U$.

PROOF. Apply Lemma 3.21 with $q = 0$, $t = 16$, $(u,v) \in \{(1,0),(1,1),(10,0)\}$.

LEMMA 3.23. If $256 \leq n \leq 361$, then $n \in U$.

PROOF. We apply Theorem 2.7 with $q = 17,19$ in (i) and $q = 16,17$ in (ii) to obtain $n \in B(15,16,17,18,19)$ for $256 \leq n \leq 307$ or $341 \leq n \leq 361$. Apply Lemma 3.19 with $q = 19$, $m = 19$ to obtain an oval of size 19 in $T(19,1;19)$. We delete a block from the $T(19,1;19)$ to obtain a $GD(\{18,19\},1,18;342)$ with an oval of size 18, and delete a block of size 19 from the GDD to obtain a $GD(\{17,18,19\},1,17;323)$ with an oval of size 16. By deleting some points from these ovals we have $n \in U$ for $308 \leq n \leq 340$, and the proof of the Lemma is complete.

LEMMA 3.24. $\{417,433,449,456,457,464,465,473,504,537\} \subset B(Q)$.

PROOF. We have $\{417,433,449,473,537\} \subset B(Q)$ from $\{52,54,56,59,67\} \subset D_1$, which will be shown in Section 4 below. The remaining values come from $\{57,58,63\} \subset U$.

COROLLARY 3.25. $E_1 = \{431,432,461,463,464,465,466,467,468,473, 474,482,487,488,491,492,493,497,498,499,500,501,504,505,506,507, 508,509, 511,516,517,518,551,552\} \subset U$.

PROOF. We apply Lemma 3.8 with the parameters shown in Table 3, where we need $N(t) \geq 7$ which comes from [3]. The condition $t+q \in B(Q)$ comes from Lemma 3.24.

LEMMA 3.26. $n \in U$ if $362 \leq n \leq 814$ and $n \notin E_2$ where $E_2 = \{383,384, 390,391,439,462,486,494,495,496,502,503,510,525,559\}$.

TABLE 3

t+u	t	u	q
431	416	15	1
432	417	15	0
461	417	44	0
463	448	15	1
464	449	15	0
465	449	16	0
466	449	17	0
467	449	18	0
468	449	19	0
473	457	16	0
474	457	17	0
482	433	49	0
487	472	15	1
488	473	15	0
491	448	43	1
492	449	43	0
493	449	44	0

t+u	t	u	q
497	448	49	1
498	449	49	0
499	449	50	0
500	449	51	0
501	457	44	0
504	455	49	1
505	456	49	1
506	457	49	0
507	457	50	0
508	457	51	0
509	465	44	0
511	503	8	1
516	465	51	0
517	473	44	0
518	503	15	1
551	536	15	1
552	503	49	1

PROOF. We apply Lemma 3.16 with $t = 43,49,64,71,73,81,83,89$ and $q \in \{0,1\}$ to get $8t+u+v+q \in U$ where the required parameters are shown in Table 2 and Table 4. We can also apply Lemma 3.15 to show that $8t+m \in U$ for $m = 3,7,12,14,21,41,42,46,48,55,56$, where for $m \in \{41,42,46,48,55,56\}$, we take $(a,b) \in \{(0,41),(8,35), (10,36),(8,41),(15,41)\}$. All the values of n have been covered except for $n \in E_1 \cup \{385\} \cup \{389,398,469,470,471,472,475, 476,481\}$, which are taken care of by Corollaries 3.25, 3.14 and

Lemma 3.15, where we take t = 43, (a,b) ∈ {(10,35),(19,35)}, t = 49, (a,b) ∈ {(43,35),(44,36), (43,41),(49,41)}.

TABLE 4

u+v+q	u	v	q	u+v+q	u	v	q
57	42	14	1	80	63	17	0
58	43	15	0	81	63	18	0
59	43	16	0	82	63	19	0
60	43	17	0	83	64	19	0
61	43	18	0	84	65	19	0
62	43	19	0	85	42	42	1
63	44	19	0	86	43	43	0
64	49	15	0	87	44	43	0
65	49	16	0	88	44	44	0
66	49	17	0	89	71	18	0
67	49	18	0	90	71	19	0
68	49	19	0	91	48	42	1
69	50	19	0	92	49	43	0
70	51	19	0	93	49	44	0
71	63	8	0	94	50	44	0
72	57	15	0	95	51	44	0
73	57	16	0	96	79	17	0
74	57	17	0	97	48	48	1
75	57	18	0	98	49	49	0
76	57	19	0	99	50	49	0
77	58	19	0	100	50	50	0
78	63	15	0	101	51	50	0
79	63	16	0	102	51	51	0

LEMMA 3.27. Suppose $N(t) \geq 17$ and $q \in \{0,1\}$, $u,v,w,s \leq t$. Then $\{t+q,u+q,v+q,w+q,s+q\} \subset U$ implies $15t+u+v+w+s+q \in U$.

PROOF. In a T(19,1;t), we delete t-u points from the first group, t-v points from the second group, t-w points from the third group and t-s points from the fourth group to obtain a {15,16,17,18,19}-GDD of group-type $t^{15}u^1v^1w^1s^1$. We then adjoin q infinite points to the groups of this GDD in order to obtain the desired result.

COROLLARY 3.28. $E_3 = \{831,832,833,834,838,839,840,847,848\} \subset U$.

PROOF. Apply Lemma 3.27 with t = 49, q = 0, s = 49, (u,v,w) ∈ {(19,19,9),(19,,19,10),(49,0,0),(49,1,0),(44,10,0),(44,10,1), (19,19,18),(44,19,0),(44,19,1)} to obtain the desired result.

LEMMA 3.29. Suppose $N(t) \geq 7$ and $q \in \{0,1\}$, $u \leq t$. Then

$$\{t+q, u+q\} \subset U \text{ implies } 8t+u+q \subset U.$$

PROOF. In a $T(9,1;t)$, we delete $t-u$ points from one group to obtain an $\{8,9\}$-GDD of group-type t^8u^1. We then adjoin q infinite points to the groups of this GDD in order to obtain the desired result.

We are now in the position to prove the following theorem.

THEOREM 3.30. $n \in U$ for every positive integer n, with the possible exception of those values listed in Table 5, where the notation "a—b" denotes the set $\{n: a \leq n \leq b\}$.

TABLE 5

2	3	4	5	6	7	11	12	13	14
"20	— 42"	45	46	47	48	52	53	54	55
56	59	60	61	62	67	68	69	70	75
76	77	78	93	94	95	96	97	101	102
103	104	105	106	107	108	109	110	111	112
115	116	117	118	119	123	124	125	126	132
133	"191	— 232"	235	236	237	238	243	244	245
246	383	384	390	391	439	462	486	494	495
496	502	503	510	525	559				

PROOF. From the previous lemmas and corollaries, we know that the conclusion holds for $n \leq 814$. Applying Lemma 3.16 with $t = 99$, $q = 0,1$ and the parameters shown in Tables 2 and 4, we know that the conclusion holds for $n \leq 894$ except $n \in E_3$, which are taken care of by Corollary 3.28. For $t \geq 99$, in Table 4 we can choose q, v and u, say $79 \leq u \leq 92$, $134 \leq u \leq 187$, or $247 \leq u \leq 361$, such that $8t+u+v+q \in U$ and $57 \leq u+v+q \leq 190$, 374, or 806, and then we obtain $n \in U$ for $849 \leq n \leq 6598$ except $1871 \leq n \leq 2104$ as illustrated in Table 6.

Apply Lemma 3.27 with $t = 113$, $q = 0,1,57 \leq u+v+q \leq 190$ as before, $w = 0$, 113, $s = 0$, 113. We find that the conclusion holds for $1752 \leq n \leq 2111$.

TABLE 6

t	$8t+57 \le n \le 8t+190$		
99	849	—	982
113	961	—	1094
128	1081	—	1214
144	1209	—	1342
157	1313	—	1446
173	1441	—	1574

t	$8t+57 \le n \le 8t+374$		
187	1553	—	1870
256	2105	—	2422
293	2401	—	2718
331	2705	—	3022
369	3009	—	3326
407	3313	—	3630

t	$8t+57 \le n \le 8t+806$		
445	3617	—	4366
538	4361	—	5110
631	5105	—	5854
724	5849	—	6598

We now apply Lemma 3.29 recursively for each $t \geq 781$ and $q = 0$, $79 \leq u \leq 92$ to get $8t+u \in U$. This guarantees that $n \in U$ whenever $n \geq 6327$, and the proof of the theorem is complete.

4. DETERMINATION OF D_1

LEMMA 4.1. $\{2,6,7,34,36,104,105,112\} \subset D_1$.

PROOF. Since $17,49,57 \subset B(8,9,17,49)$, it is clear that $2,6,7 \in D_1$. $34,36 \in D_1$ come from $273,289 \in B(17)$. $104 \subset D_1$ comes from $T(17,1;\ 49)$ and $833 \in B(Q)$. $105,120 \in D_1$ come from $120,128 \in B(Q) \subset R_8$ and $841,897 \in B(8)$.

LEMMA 4.2. If there exists a $GD(U,1,D_1;d)$, then $d \in D_1$.

PROOF. In the $GD(U,1,D_1;d)$ we give weight 8 to every point. Since a $GD(B(Q),1,8;8u)$ exists for every block of size $u \in U$, Construction 2.14 gives us a GDD with block sizes all in $B(Q)$ and group sizes of the form $8k$ where $k \in D_1$. We then adjoin one infinite point to this GDD to obtain an $(8d+1,B(Q),1)$-PBD, and the result follows.

COROLLARY 4.3. $\{96,97\} \subset D_1$.

PROOF. We shall apply Lemma 4.2 as follows. In a $T(10,1;11)$, we delete one block and some points in a group in such a way that the small group is of size 6 or 7, then we get 96 or 97 $\in D_1$.

LEMMA 4.4. If $N(t) \geq 7$ and $t+q \in B(Q)$ where $q = 0$ or 1, then $t+e \in D_1$ provided that $e \in D_1$ and $8e+1 \leq t+q$.

PROOF. In a $T(9,1;t)$ we delete some points from a group so that the group size becomes $8e+1-q$. We then adjoin q infinite points to each group of the resulting GDD to get a $(8t+8e+1,\{8,9,t+q,8e+1\},1)$-PBD and $t+e \in D_1$.

COROLLARY 4.5. $\{55,59,62,67,75,123\} \subset D_1$.

PROOF. We apply Lemma 4.4 with the parameters shown in Table 7.

TABLE 7

t+e	t	e	q
55	49	6	0
59	57	2	0
62	56	6	1
67	65	2	0
75	73	2	0
123	121	2	0

LEMMA 4.6. Suppose $N(t) \geq 8$. Then $\{t,u,v\} \subset D_1$ implies $8t+u+v \in D_1$, where $u,v \leq t$.

PROOF. In a $T(10,1;t)$, we delete t-u points from one group and t-v points from another group to obtain an $\{8,9,10\}$-GDD of group-type $t^8u^1v^1$. Apply Lemma 4.2 with the fact that $\{t,u,v\} \subset D_1$ to obtain the desired result.

COROLLARY 4.7. $\{76,384,494,495,496,502,503,510,525,559\} \subset D_1$.

PROOF. We apply Lemma 4.6 with the parameters shown in Table 8.

LEMMA 4.8. Suppose $N(t) \geq 15$. Then $\{t,u,v\} \subset D_1$ implies $15t+u+v \in D_1$, where $u,v \leq t$.

TABLE 8

8t+u+v	t	u	v
76	9	2	2
384	43	34	6
494	59	15	7
495	59	16	7
496	59	16	8
502	59	15	15
503	59	16	15
510	59	19	19
525	59	34	19
559	59	44	43

PROOF. In a T(17,1;t), we delete t-u points from one group and t-v points from another group to obtain a {15,16,17}-GDD of group-type $t^{15}u^1v^1$. Apply Lemma 4.2 with the fact $\{t,u,v\} \subset D_1$ to obtain the desired result.

COROLLARY 4.9. $\{243,244\} \subset D_1$.

PROOF. Apply Lemma 4.8 with t = 16, u = 2, v = 1 or 2 to obtain the desired result.

LEMMA 4.10. If $v \in RB(8)$, then $v/8+e \in D_1$ provided that $e \in D_1$ and $8e+1 \leq (v-1)/7$.

PROOF. We adjoin 8e+1 infinite points to a resolvable (v,8,1)-BIBD, where one infinite point is adjoined to each of 8e+1 parallel classes of blocks. We obtain $v+8e+1 \in B(8,9,8e+1) \subset B(Q)$ and the conclusion follows.

COROLLARY 4.11. $\{45,52,56,70,77,78,133,245,246\} \subset D_1$.

PROOF. There is a resolvable (v,8,1)-BIBD with v = 344,400,512, 568,1016, or 1912 as shown in Lemma 3.6. We then apply Lemma 4.10 with e = 2, 6 or 7 to get the desired result.

LEMMA 4.12. If $t \in B(Q)$ is a prime power, then $t+e \in D_1$ provided $e+1 \in D_1$ and $8e+8 \leq t$.

PROOF. In a T(t+1,1;t), we delete all the points in t-8 groups except for 8e+1 points which lie in the same block. This gives

$8t+8e+1 \in B(8,9,t,8e+9) \subset B(Q)$ and $t+e \in D_1$.

COROLLARY 4.13. $\{54,69,126,439,462,486\} \subset D_1$.

PROOF. We apply lemma 4.12 with $t \in \{49,64,121,433,457\}$ and $e \in \{5,6,53\}$.

The following lemma is a more general form of Lemma 4.4.

LEMMA 4.14. Suppose $t+q \in B(Q \cup \{q^*\})$ and $N(t) \geq 7$. Then $t+e \in D_1$ provided $e \in D_1$ and $8e+1-q \leq t$.

PROOF. In a $T(9,1;t)$, we delete some points from one group so that the truncated group size is $8e+1-q$. We then adjoin q infinite points to the groups of the resulting GDD so as to form an $(8t+8e+1,B(Q),1)$-PBD and get $t+e \in D_1$.

COROLLARY 4.15. $\{115,118,119,390,391\} \subset D_1$.

PROOF. We apply Lemma 4.14 with $q = 8$ and $(t,e) \in \{(112,6), (112,7),(113,2),(384,6),(384,7)\}$ to obtain the desired result. The conditions $120 \in B(8)$, $121 \in B(8,9)$ come from the proof of Corollary 3.14, and $392 \in B(8,49)$ from $49 \in Q$.

LEMMA 4.16. There exist {8,9}-GDDs of the following group-types:

(a) 7^8, (b) 7^9, (c) $7^8 8^1$, (d) $7^9 8^1$.

PROOF. (a), (b) are fairly obvious. For (c) we adjoin one infinite point to each group of a $T(8,1;8)$, and then delete one point from a block of size 9. For (d), we delete one point from a $T(8,1;9)$.

LEMMA 4.17. Suppose $N(t) \geq 8$, and $\{7t+q,7u+q\} \subset B(Q \cup \{q^*\})$, $u \leq t$. Then $56t+7u+8v+q \in B(Q)$ provided $8v+q \in B(Q)$ and $v \leq t$.

PROOF. In all groups but the first two of a $T(10,1;t)$, give the points weight 7. In the first group, we give u points weight 7 and the remaining points weight 0. In the second group, we give v points weight 8 and give the remaining points weight 0. We can apply Construction 2.14 with the necessary input designs from Lemma 4.16 to obtain a {8,9}-GDD of group-type $(7t)^8(7u)^1(8v)^1$. We then adjoin a set of q infinite points to the groups of this GDD, using the fact that $\{7t+q, 7u+q\} \subset B(Q \cup \{q^*\})$ and $8u+q \in B(Q)$ to obtain the desired result.

COROLLARY 4.18. $132 \in D_1$.

PROOF. We apply Lemma 4.17 with t = 17, u = 8, v = 6 and q = 1 to obtain $1057 \in B(Q)$ and $132 \in D_1$.

Summarizing the above result, we have proved

THEOREM 4.19. $n \in D_1$ for every positive integer n, with the possible exception of those values listed in Table 9, where the notation "a—b" denotes the set $\{n: a \leq n \leq b\}$.

TABLE 9

3	4	5	11	12	13	14	"20 — 33"	35	37	38
39	40	41	42	46	47	48	53	60	61	68	93
94	95	101	102	103	106	107	108	109	110	111	116
117	124	125	"191 — 232"	235	236	237	238	383		

5. DETERMINATION OF D_0

LEMMA 5.1. $106 \in D_0$.

PROOF. Since $121 \in R_8$, so $848 \in B(8)$ and $106 \in D_0$.

LEMMA 5.2. $\{70,77,78,126,133\} \subset D_0$.

PROOF. We apply Lemma 4.17 with the parameters shown in Table 10.

TABLE 10

d	8d	t	7u	8v	q
70	560	9	7	48	1
77	616	9	63	48	1
78	624	9	63	56	1
126	1008	17	7	48	1
133	1064	17	63	48	1

LEMMA 5.3. Suppose $N(t) \geq 7$ and $t+q \in B(Q \cup \{q^*\})$. Then $8t+u+q \in B(Q)$ provided $u+q \in B(Q)$ and $u \leq t$.

PROOF. In a $T(9,1;t)$, we delete $t-u$ points from one group, and then adjoin q infinite points to the groups of the resulting GDD so as to form an $(8t+u+q,B(Q),1)$-PBD.

COROLLARY 5.4. $439 \in D_0$.

PROOF. We apply Lemma 5.3 with $t = 424$, $u = 111$, $q = 9$. The condition $433 \in B(Q \cup \{q^*\})$ come from the proof of Corollary 4.13.

LEMMA 5.5. There exist $\{8,9\}$-GDDs of the following group-types:

(a) 8^8, (b) 8^9, (c) 8^{10}, (d) $8^8 7^1$, (e) $8^9 7^1$.

PROOF. (a), (b) and (c) are fairly obvious. For (d), we delete one point from a $T(9,1;8)$. For (e), we take an $RT(8,1;9)$ and adjoin 7 points to 7 parallel classes.

LEMMA 5.6. Suppose $N(t) \geq 8$ and $\{8t+q, 8x+7y+q\} \subset B(Q \cup \{q^*\})$, $x+y \leq t$. Then $64t+8x+7y+8u+q \in B(Q)$ provided $8u+q \in B(Q)$ and $u \leq t$.

PROOF. In all groups but two of a $T(10,1;t)$, we give the points weight 8. In the second last group, we give u points weight 8 and give the remaining points weight 0. In the last group, we give weight 8 to x points, weight 7 to y points and give the remaining points weight 0. We can apply Construction 2.14 with the necessary input designs from Lemma 5.5 to obtain an $\{8,9\}$-GDD of group-type $(8t)^8(8u)^1(8x+7y)^1$. We then adjoin a set of q infinite points to the groups of this GDD, using the fact that $\{8t+q,8x+7y+q\} \subset B(Q \cup \{q^*\})$ and $8u+q \in B(Q)$ to obtain the desired result.

COROLLARY 5.7. $\{75,390,486,494,495,496,502,503,510,525,559\} \subset D_0$.

PROOF. We apply Lemma 5.6 with the parameters shown in Table 11. The conditions 345, 352 $\in B(Q \cup \{q^*\})$ come from 344 $\in$ RB(8).

LEMMA 5.8. There exist {8,9,17}-GDDs of the following group-types:

(a) 8^{15}, (b) 8^{16}, (c) 8^{17}, (d) $8^{15}7^1$, (e) $8^{16},7^1$.

TABLE 11

d	8d	t	8x	7y	8u	q
75	600	9	0	7	16	1
390	3120	43	72	7	288	1
486	3888	59	56	7	48	1
494	3952	59	112	7	56	1
495	3960	59	112	7	64	1
496	3968	59	112	7	72	1
502	4016	59	112	7	120	1
503	4024	59	112	7	128	1
510	4080	53	336	7	336	9
525	4200	59	56	7	360	1
559	4472	59	344	7	344	1

PROOF. (a), (b) and (c) are fairly obvious. For (d), we take a resolvable (120,8,1)-BIBD and adjoin 7 points to 7 parallel classes. For (e), we adjoin a point to a T(9,1;16) to obtain a (145,{9,17},1)-PBD, and then delete 10 points from the block of size 17.

LEMMA 5.9. Suppose $N(t) \geq 15$, and $\{8t+q, 8x+7y+q\} \subset B(Q \cup \{q^*\})$, $x+y \leq t$. Then $120t+8x+7y+8u+q \in B(Q)$ provided $8u+q \in B(Q)$ and $u \leq t$.

PROOF. In all groups but two of a T(17,1;t), we give the points weight 8. In the second last group, we give u points weight 8 and give the remaining points weight 0. In the last group, we give weight 8 to x points, weight 7 to y points and give the remaining points weight 0, we can apply Construction 2.14 with the necessary input designs from Lemma 5.8 to obtain a {8,9,17}-GDD of group-type $(8t)^{15}(8u)^1(8x+7y)^1$. We then adjoin a set of q infinite points to the groups of this GDD, using the fact that $\{8t+q, 8x+7y+q\} \subset B(Q \cup \{q^*\})$ and $8u+q \in B(Q)$ to obtain the desired result.

COROLLARY 5.10. $243 \in D_0$.

PROOF. We apply Lemma 5.9 with $t = 16$, $8u = 16$, $8x = 0$, $7y = 7$, $q = 1$ to obtain $1944 \in B(Q)$ and $243 \in D_0$.

Combining Theorem 3.30 and the results of this section, we have

THEOREM 5.11. $n \in D_0$ for every positive integer n, with the possible exception of those values not underlined in Table 5.

6. CONCLUSION

LEMMA 6.1. $\{385, 448, 3696\} \subset R_8$.

PROOF. In a $T(8,1;\ 336)$, we adjoin a set of 8 infinite points to each group and using the fact $344 \in B(8)$, to get $2696 \in B(8,344) \subset B(8)$ and $385 \in R_8$. $448 \in R_8$ come from $3137 \in B(8)$ (see [12]). In a $T(9,1;\ 447)$, we delete 328 points from one group and then adjoin one infinite point to the groups of the resulting GDD, to get $3696 \in B(8,9,120,448) \subset R_8$.

PROOF OF THEOREM 1.4. The result is an immediate consequence of Theorems 4.19 and 5.11.

PROOF OF THEOREM 1.2. Since $B(Q) \subset R_8$, the result is an immediate consequence of Theorem 1.4 and Lemma 6.1.

REMARK. After this paper was written, we saw the paper by R. Mathon and A. Rosa, "Tables of parameters of BIBDs with $r \leq 41$ including existence, enumeration, and resolvability results," Ann. Discrete Math., 26(1985), 275-308, in which the existence of a (232,8,1)-BIBD is indicated. This results in the deletion of 33 and some other numbers from our Table 1.

REFERENCES

[1] Th. Beth, D. Jungnickel and H. Lenz, Design Theory, Bibliographisches Institut, Zurich, 1985.

[2] R.C. Bose, On the application of finite projective geometry for deriving a certain series of balanced Kirkman arrangements, Calcutta Math. Soc. Golden Jubilee Vol., 1959, 34-354.

[3] A.E. Brouwer, The number of mutually orthogonal Latin squares -- a table up to order 10000, Research Report ZW123/79, Mathematisch Centrum, Amsterdam, 1979.

[4] A.E. Brouwer and G.H.J. Van Rees, More mutually orthogonal Latin squares, Discrete Math., 39(1982), 263-281.

[5] M. Hall, Jr., Combinatorial Theory, Second Edition, John Wiley & Sons, New York, 1986.

[6] H. Hanani, Balanced incomplete block designs and related designs, Discrete Math. 11(1975), 255-369.

[7] H. Hanani, D.K. Ray-Chaudhuri and R.M. Wilson, On resolvable designs, Discrete Math. 3(1972), 343-357.

[8] H.F. MacNeish, Euler squares, Ann. Math. 23(1922), 221-227.

[9] R. Mathon, On the existence of doubly resolvable Kirkman systems and equidistant permutation arrays, Discrete Math. 30(1980), 157-172.

[10] W.H. Mills, Balanced incomplete block designs with $k = 6$ and $\lambda = 1$, Enumeration and Design, (Ed. D.M. Jackson & S.A. Vanstone) Academic Press, Canada, 1984, 239-244.

[11] R.C. Mullin, D.G. Hoffman and C.C. Lindner, A few more BIBDs with $k = 6$ and $\lambda = 1$, Ann. Discrete Math. 34(1987), 379-384.

[12] R.C. Mullin, Finite bases for some PBD-closed sets, preprint.

[13] E. Seiden, A method of construction of resolvable BIBD, Sankhya (A) 25 (1963), 393-394.

[14] J.H. Van Lint, Combinatorial Theory Seminar, Eindhoven University of Technology. Lecture Notes in Mathematics 382. Springer, Berlin-Heidelberg-New York, 1974.

[15] R.M. Wilson, Cyclotomy and difference families in elementary Abelian groups, J. Number Theory 4(1972), 17-47.

[16] R.M. Wilson, Constructions and uses of pairwise balanced designs, Mathematical Centre Tracts 55(1974), 18-41.

[17] Zhu Lie, Du Beiliang and Yin Jianxing, Some new balanced incomplete block designs with k = 6 and $\lambda = 1$, Ars Combinatoria 24(1987), 167-174.

Partially Resolvable 3-(3,4,v:m) Partitions

GUO Haitao, Department of Applied Mathematics, Shanghai Jiao Tong University, Shanghai, P.R. China

1. INTRODUCTION

An ordered pair $(X,\mathcal{B})$ called a t-(v,K,λ)-partition is defined as a set X of v points together with a collection $\mathcal{B}$ of k_i-subsets on X, called blocks, with $k_i \in K$, such that every t-subset of X is contained in exactly λ blocks of X.

Given a t-(v,K,λ)-partition $(X,\mathcal{B})$, a parallel class is a subset of $\mathcal{B}$ which partitions X. A partially resolvable t-partition $(X,\mathcal{P},\mathcal{B})$, PRP t-$(P,S,v;M)$, is defined to be a t-$(v,K,1)$ partition such that:

a) $K = P \cup S, \quad P \cap S = \emptyset$

b) $|P| = |M|$

c) If $P = \{p_1,p_2,\ldots,p_n\}$ and $M = \{m_1,m_2,\ldots,m_n\}$, the blocks of size p_i can be partitioned into m_i parallel classes. Here $t \geq 2$ and $p,s \geq t$ for every $p \in P$, $s \in S$.

$\mathcal{P}$ is a collection of parallel classes of blocks of size p with $p \in P$ and $\mathcal{B}$ is a collection of blocks of size s with $s \in S$.

A partially resolvable partition $(X_1,\mathcal{P}_1,\mathcal{B}_1)$ of order V has a subsystem $(X_2,\mathcal{P}_2,\mathcal{B}_2)$ of order v if there exist sets $X_2 \subseteq X_1$,

$\mathcal{P}_2 \subseteq \mathcal{P}_1$ such that $(X_2, \mathcal{P}_2, \mathcal{B}_2)$ is a partially resolvable partition of order v.

Generally we are concerned primarily with the case $|P| = 1$, $|S| = 1$. We will denote the corresponding partially resolvable t-partition by PRP t-(p,s,v;m) provided P = {p}, S = {s}.

Huang, Mendelsohn and Rosa [3] determined the existence of PRP 2-(p,s,v;m) in some simple cases; specifically, the necessary conditions for the existence of a PRP 2-(2,3,v;m) are also sufficient and the necessary conditions for the existence of a PRP 2-(3,2,v;m) are sufficient except for the cases (v,m) = (6,2),(12,5). For a PRP 2-(2,4,v;m), they conjectured that the necessary conditions for the existence of a PRP 2-(2,4,v;m) are sufficient except for (v,m) = (8,1),(10,3),(12,2). When considering embeddings of BIBDs of block size four, Rees and Stinson [4] proved that the necessary conditions for the existence of a PRP 2-(3,4,v;m) are sufficient. To construct a PRP t-(p,s,v;m) with $t \geq 3$ is not easy. This paper deals with the existence of a PRP 3-(4,3,v;m) and obtains some recursive constructions.

2. A NECESSARY CONDITION

Observe that our definition allows a PRP to have blocks of size p only or to have blocks of size s only. In the former case a PRP t-(p,0,v;m) is a resolvable Steiner system RS(t,p,v) while in the latter case m = 0, a PRP t-(0,s,v;0) is a Steiner system S(t,s,v).

In general, we have

$$0 \leq m \leq \binom{v-1}{t-1} / \binom{p-1}{t-1},$$

because, for a given element, a PRP t-(p,s,v;m) in which there are $\binom{v-1}{t-1}$ t-subsets containing that element has m parallel classes of blocks and every parallel class has $\binom{p-1}{t-1}$ t-subsets containing the element. We have the following necessary conditions for the existence of a PRP t-(p,s,v;m):

$$v \equiv 0 \pmod{p} \text{ if } m \geq 0 \qquad (1)$$

$$\binom{s}{t} \,\Big|\, \left[\binom{v}{t} - mv\binom{p}{t}/p\right] \qquad (2)$$

$$\binom{s-1}{t-1} \mid [\binom{v-1}{t-1} - m\binom{p-1}{t-1}] \tag{3}$$

Condition (1) is obvious: if there is at least one parallel class of blocks of size p then p must divide v. Conditions (2) and (3) are obtained by counting the total number of blocks of size s and those containing a given element respectively.

Now we consider the case of a PRP 3-(4,3,v;m). In the case where there are no blocks of size 3, it is a resolvable Steiner quadruple system RSQS(3,4,v). Hartman [2] proved that the necessary condition for the existence of a RSQS(3,4,v), $v \equiv 4$ or 8 (mod 12), is also sufficient with the possible exception of twenty-three values of v. In the case m = 0, a PRP is a Steiner system S(3,3,v), which is trivial for any v.

In general, by conditions (1), (2) and (3), the necessary conditions for the existence of a PRP 3-(4,3,v;m) are $v \equiv 0$, 4 or 8 (mod 12), $0 \leq m \leq (v-1)(v-2)/6$.

3. DEFINITIONS AND PRELIMINARY LEMMA

A one-factor F of the complete graph on X is a set of 2-subsets of X, called edges, which partitions X. A one-factorization of the complete graph on X is a partition $F = F_1 \cup F_2 \cup \ldots \cup F_{v-1}$ of the set of all 2-subsets of X into one-factors.

The number $|X|$ of vertices of a complete graph on X with a one-factorization must be even, since the number of the edges in each one factor is $|X|/2$.

Throughout the sequel, Z_n will denote the cyclic group of integers modulo n under addition. I_n will denote the set of integers $\{1,2,\ldots,n\}$.

For $x \in Z_n$, we define $|x|$ by

$$|x| = \begin{cases} x & \text{if} \quad 0 \leq x \leq n/2 \\ -x & \text{if} \quad n/2 < x < n-1. \end{cases}$$

For $n \geq 2$ and $L \subseteq I_{\lceil n/2 \rceil}$, here $\lceil n/2 \rceil$ equals n/2 if n is even or (n+1)/2 if n is odd, G(n,L) is defined to be the regular graph with vertex set Z_n and edge set E given by $(x,y) \in E$ if and only if $|x-y| \in L$.

A resolvable pairing RP(2k,s) is defined as an ordered set $(\Delta, R_i, PR_i, S_i, PS_i, \overline{R}_i)$ which satisfies the following conditions for each $i \in Z_3$:

(1) RP(2k,s) consists of the subsets Δ, R_i, S_i, $\overline{R}_i$ of Z_{12k+s}. The members of sets PR_i and PS_i are pairs of the elements from Z_{12k+s}.

(2) Cardinality and Symmetry Conditions

a) $|\Delta| = s$, $|R_i| = |S_i| = 4k$, $|\overline{R}_i| = 2k$

b) $x \in \Delta$ iff $-x \in \Delta$

(3) Partitioning Conditions

a) PR_i and PS_i are partitions of R_i and S_i respectively into pairs

b) $Z_{12k+s} = \Delta/R_0/R_1/R_2 = \Delta/R_i/S_i/\overline{R}_{i-1}/\overline{R}_{i+1}$

(4) Pairing Conditions

Let: $L_i = \{|x-y| : (x,y) \in PR_i\}$

and $O_i = \{|x-y| : (x,y) \in PS_i\}$

a) No maximum length chords

$$(12k+s)/2 \notin L_i \cup O_i$$

b) Distinct and odd chords

$$|L_i| = |O_i| = 2k, \quad L_i \cap O_i = \emptyset$$

and all members of O_i are odd

c) One-factorization

The complement G_i of the graph $G(12k+s, L_i \cup O_i)$ has a one-factorization

Hartman [1] proved the following lemma:

LEMMA 1. Resolvable pairings RP(2k,s) exist for all integers $k \geq 0$ and $s > 2$, s even.

4. MAIN RESULT

LEMMA 2. If there exists a PRP 3-(4,3,v;m), then there exists a PRP 3-(4,3,v;m′), where $m' \leq m$.

PROOF. In a PRP 3-(4,3,v;m), delete m-m′ parallel classes add all triples contained in the deleted parallel classes. Then a

PRP 3-(4,3,v;m′) results.

THEOREM 1. Let $v \equiv 4$ or $8 \pmod{12}$. Then there exists a PRP 3-(4,3,v;m), whenever $m \leq (v-1)(v-2)/6$ with the possible exception of twenty-three values.

PROOF. When $v \equiv 4$ or $8 \pmod{12}$, Hartman [2] proved that there exists an RSQS(3,4,v) with the possible exception of twenty-three values of v. The number of parallel classes in an RSQS(3,4,v) is $(v-1)(v-2)/6$. By Lemma 2, the theorem is proved.

THEOREM 2. If there exists a PRP 3-(4,3,v;m), then there exists a PRP 3-(4,3,2v;m′) containing a subsystem PRP 3-(4,3,v;m), whenever $m \leq m' \leq m+v(v-1)/2$.

PROOF. Let $(X,\mathcal{P},\mathcal{B})$ be a PRP 3-(4,3,v;m) on the point set I_v, where $p_i \in \mathcal{P}$ is a parallel class for $i \in I_m$. A PRP 3-(4,3,2v; m+v(v-1)/2) can be constructed on the point set $I_v \times I_2$. The parallel classes will be constructed as follows: The first m parallel classes are

$$P(j) = p_j\times\{1\} \cup p_j\times\{2\} \quad \text{for} \quad 1 \leq j \leq m.$$

It is easy to find a one-factorization $F = F_1 \cup F_2 \cup \ldots \cup F_{v-1}$ of the complete graph on X. Let (e_{mn}) be a Latin square of side v/2. Each of v(v-1)/2 further parallel classes P′(i,j) comprise v/2 blocks

$$(a,b,c,d)$$

where (a,b) is the k-th edge of $F_i\times\{1\}$ and (c,d) is the e_{kj}-th edge of $F_i\times\{2\}$. Finally, all triples not contained in the above parallel classes are added. By Lemma 2, the proof is completed.

By the tripling construction given by Hartman [1], we can obtain the following theorem:

THEOREM 3. If $V \equiv 2v \pmod{12}$ and there exists a PRP 3-(4,3,V;M+m) with a subsystem PRP 3-(4,3,v;m), then there exists a PRP 3-(4,3,3V-2v;M′+m) with subsystems PRP 3-(4,3,v;m) and PRP 3-(4,3,V;M+m), where $r(v) = (v-1)(v-2)/6$, $M = r(V)-r(v)$, $M' = r(3V-2v)-r(v)$.

PROOF. Since $V \equiv 2v \pmod{12}$, we can write $V = 12k+2v$, where $k \geq 0$. By Lemma 1, with $i \in Z_3$, there is an RP(2k,v); let $(\Delta, R_i, \overline{R}_i, S_i, PR_i, PS_i)$ be such a resolvable pairing. Let $F_i^{(2n-1)} \cup F_i^{(2n)}$ be a one-factorization of the graph $G(12k+v, \{m\})$, where m is the n-th member of O_i and let $F_i^{(4k+1)} \cup \ldots \cup F_i^{(8k+v-1)}$ be a one-factorization of G_i.

Let $(X_1, \mathcal{P}_1, \mathcal{B}_1)$ be a PRP 3-(4,3,V;M+m) with a sub-PRP 3-(4,3,v;m) $(X_2, \mathcal{P}_2, \mathcal{B}_2)$ where $X_1 \supseteq X_2$ and $\mathcal{P}_1 = \bigcup_{i=I_{M+m}} P_{1,i}$, $\mathcal{P}_2 = \bigcup_{j \in I_m} P_{2,j}$, $P_{1,j} \supseteq P_{2,j}$, where the parallel classes $P_{1,i}$ and $P_{2,j}$ satisfy $j \in I_m$. Let

$$X' = X_2 \cup \{a_i : a \in X_1 \backslash X_2,\ i \in Z_3\}$$

and for each $i \in Z_3$, define an embedding $\lambda : X_1 \to X'$ by

$$\lambda_i a = \begin{cases} a & \text{if } a \in X_2 \\ a_i & \text{if } a \in X_1 \backslash X_2. \end{cases}$$

Now we construct a PRP 3-(4,3,3V-2v;M′+m) on the point set X′ as follows, using the one-factorization $F_i^{(h)}$ given above and a Latin square (e_{mn}) of side $(12k+v)/2$.

The first m parallel classes are

$$P'(j) = \lambda_0 P_{1,j} \cup \lambda_1 P_{1,j} \cup \lambda_2 P_{1,j} \quad \text{for} \quad 1 \leq j \leq m.$$

For $m < j \leq M+m$ and $i \in Z_3$, each of the next 3M parallel classes $P'(i,j)$ comprises $V/4$ blocks from $\lambda_i P_{1,j}$ and $(V-v)/2$ blocks $(a_{i+1}, b_{i+1}, c_{i+2}, d_{i+2})$, formed by taking (a,b) the m-th edge in $F_{i+1}^{(h)}$ and (c,d) the e_{mn}-th edge of $F_{i+2}^{(h)}$, where $4k+1 \leq h \leq 8k+v-1$ and $1 \leq m,n \leq (V-v)/2$ with $h \equiv j \pmod{4k+v-1}$ and $n \equiv j \pmod{(V-v)/2}$.

The final $(V-v)^2$ parallel classes are denoted $P'(a,b,c)$ where $a+b+c \equiv 0 \pmod{V-v}$. $P'(a,b,c)$ comprises:

v blocks

$(\infty_j, (a+s)_0, (b-s)_1, (c+s)_2)$, where $\infty_j \in X_2$ and s is the j-th member of Δ;

6k blocks

$((a+s)_0,(a+t)_0,(b-u)_1,(c+u)_2)$ for $i = 0$

$((a+u)_0,(b+s)_1,(b+t)_1,(c-u)_2)$ for $i = 1$

$((a-u)_0,(b+u)_1,(c+s)_2,(c+t)_2)$ for $i = 2$

where (s,t) is the j-th edge in PR_i and u is the j-th member of $\overline{R}_i$;

2k blocks

$((a+s)_0,(a+s')_0,(b+t)_1,(b+t')_1)$

$((b+u)_1,(b+u')_1,(c+w)_2,(c+w')_2)$

$((c+z)_2,(c+z')_2,(a+y)_0,(a+y')_0)$,

where (s,s'), (y,y'), (t,t'), (u,u'), (w,w'), (z,z') are the j-th edges selected from the sets PA_i and PB_i which are partitions of PS_i into parts of size k.

Adding all triples not contained in above parallel classes. The proof is completed.

REMARK. In the proof of this theorem, it is easy to see that a PRP 3-(4,3,3V-2v;M'+m-3m') exists when $M = r(V)-r(v)-m'$.

To conclude, we present some concrete small v cases.

LEMMA 3. There exists a PRP 3-(4,3,12,13); when $m > 13$, no PRP 3-(4,,3,12;m) exists.

PROOF. First we construct a PRP 3-(4,3,12;13) on the point set $I_4 \times I_3$. I_4 has a one-factorization $F = F_1 \cup F_2 \cup F_3$. For every F_i, four parallel classes can be constructed as follows:

(a_1^1,a_1^2) (a_2^2,a_1^3) (a_2^3,a_2^1)

(a_1^1,a_2^2) (a_1^2,a_2^3) (a_1^3,a_2^1)

(a_1^1,a_1^3) (a_2^3,a_2^2) (a_2^1,a_1^2)

(a_1^1,a_2^3) (a_1^3,a_1^2) (a_2^1,a_2^2)

Here $a_j^k \in F_i \times \{k\}$, $i,k \in I_3$, $j \in I_2$. The last class is $(a,b,c,d)_i$, $a,b,c,d \in I_4$. Adding all triples not contained in the above parallel classes we get a PRP 3-(4,3,12;13).

Now we prove 13 is the largest number of parallel classes

in PRP 3-(4,3,12;m). Suppose a PRP 3-(4,3,12;m) is constructed on the points set of $A = \{a,b,c,d\} \times I_3$. We can assume the first parallel class has the form:

$$(a_1,b_1,c_1,d_1) \quad (a_2,b_2,c_2,d_2) \quad (a_3,b_3,c_3,d_3).$$

Then all blocks in the remaining parallel classes must contain a one-factor of $A\times\{i\}$ which we denote e_j. As every one-factor e_j appears at most four times in the remaining parallel classes and $A\times\{i\}$ only has three one-factors, the remaining parallel classes have at most twelve ones. So the Lemma is proved.

REMARK. Using one-factorizations, we can also construct a PRP 3-(4,3,36;13+12×11) with a subsystem PRP 3-(4,3,12;13).

COROLLARY 1. There exists a PRP 3-(4,3,v;m), where $v = 2^n \times 12$ and $m \leq r(v)-5$.

PROOF. This clearly follows from Lemma 3 and Theorem 2.

Table 1

1,	2,	4,	5,	8,	10,	11,	13,	14,	16,	20,	22,	23,	25,	26,	28,	29,
31,	32,	35,	37,	38,	40,	41,	44,	46,	47,	50,	53,	56,	58,	59,	61,	
62,	64,	65,	67,	68,	70,	71,	74,	75,	76,	77,	79,	80,	81,	82,	83,	
84,	85,	86,	87,	89,	91,	94,	95,	97,	100.							

COROLLARY 2. There exists a PRP 3-(4,3,v;m), where $v = 2^n \times 3 \times 12$ and $m \leq r(v)-53$.

By Theorem 3 and Corollary 1, for $v = 12 \times 2n$, we know a PRP 3-(4,3,v;m) exists, where $m \leq r(v)-5$, n is an integer in Table 1 and $n \leq 100$.

By Theorem 3 and Lemma 3, we can construct a PRP 3-(4,3, 7×12; r(7×12)-3×53).

ACKNOWLEDGEMENT. I wish to thank Professor Shen Hao for instruction in my research and Dr. F.E. Bennett for helpful suggestions at the conference.

REFERENCES

[1] A. Hartman, Tripling quadruple systems, Ars Combinatoria 10(1980), 255-309.

[2] A. Hartman, The existence of resolvable Steiner quadruple systems, J. Combinatorial Theory 44A(1987).

[3] C. Huang, E. Mendelsohn and A. Rosa, On partially resolvable t-partitions, Discrete Math. 12(1982), 169-183.

[4] R. Rees and D.R. Stinson, On the existence of incomplete designs of block size four having one hole, to appear.

Symmetric Mendelsohn Triple Systems and Large Sets of Disjoint Mendelsohn Triple Systems

KANG Qing-de, Mathematics Department, Hebei Normal College, China

CHANG Yan-xun, Mathematics Department, Hebei Normal College, China

Let x,y,z be distinct elements of a set S. $\langle x,y,z\rangle$ is called a *cyclic triple* on S; it contains three ordered pairs (x,y), (y,z) and (z,x). The cyclic triples $\langle x,y,z\rangle$, $\langle y,z,x\rangle$ and $\langle z,x,y\rangle$ will be regarded as identical. A *Mendelsohn triple system* of order v (MTS(v)) is a pair $(S,\mathcal{B})$, where S is a set containing v elements and $\mathcal{B}$ is a collection of cyclic triples of elements of S such that every ordered pair of distinct elements of S belongs to exactly one cyclic triple of $\mathcal{B}$. Therefore $|\mathcal{B}| = \frac{v}{3}(v-1)$. In [1] it was proven that an MTS(v) exists if and only if $v \not\equiv 2 \pmod 3$, $v > 1$, except $v = 6$.

For a MTS(v) $(S,\mathcal{B})$, if there exist $a,b \in S$ $(a \neq b)$ such that $\langle a,b,x\rangle \in B \Leftrightarrow \langle b,a,x\rangle \in \mathcal{B}$ and $\langle a,x,y\rangle \in \mathcal{B} \Leftrightarrow \langle b,y,x\rangle \in \mathcal{B}$ $(x,y \in S\setminus\{a,b\})$ then we call the MTS(v) a *symmetric Mendelsohn triple system* of order v and denote it by SMTS(v).

Two MTS(v) on the same set are said to be *disjoint* if they have no cyclic triples in common. We denote by $D_M(v)$ the maximum number of pairwise disjoint MTS(v). It is easy to see that $D_M(v) \leq v-2$. If $D_M(v) = v-2$, we call any set of v-2 pairwise disjoint MTS(v) a *large set* of disjoint MTS(v) and denote it by LMTS(v). Similarly, we can define LSMTS(v) also.

1. LSMTS(v) OF ODD ORDER v

For $v \equiv 1,3 \pmod 6$, $v > 1$, let D(v) be the maximum number of pairwise disjoint STS(v). Then $D(v) \leq D_M(v)$. It is well known that D(7) = 2, but in [4] it is shown $D_M(7) = 5$. For $v \geq 9$, $v \equiv 1$ or $3 \pmod 6$, $v \notin T = \{141,283,501,789,1501,2365\}$ we have $D(v) = v-2$ ([3]). Therefore, for $v \equiv 1$ or $3 \pmod 6$, $v > 1$, $v \notin T$, we already have $D_M(v) = v-2$. Thus for the existence of LMTS of odd order only the six numbers in T remain unsettled. Now we shall give up this approach and construct directly all LMTS(v) of odd order v.

THEOREM 1. If $v \equiv 1$ or $3 \pmod 6$, $v > 1$, then there exists an LSMTS(v).

CONSTRUCTION. Let $u \equiv \pm 1 \pmod 6$, $S = \{a,b\} \cup Z_u$, where $Z_u = \{0,1,\ldots,u-1\}$ is the ring of residues modulo u, $a,b \notin Z_u$ and $v = u+2 = |S|$. We can construct u cyclic triple systems $\mathcal{B}_i$ $(i \in Z_u)$ on S: given i, $\mathcal{B}_i$ contains:

1) $\langle a,b,i\rangle$, $\langle b,a,i\rangle$;

2) $\langle a,x+i,-2x+i\rangle$, $\langle b,-2x+i,x+i\rangle$, where $x \in Z_u\backslash\{0\}$ (this gives 2(u-1) cyclic triples);

3) $\langle x,y,z\rangle$, $\langle y,x,z\rangle$, where each $\{x,y,z\}$ is an unordered 3-subset of Z_u such that $x+y+z \equiv 3i \pmod u$ (this gives $\frac{2}{u}\binom{u}{3}$ cyclic triples).

PROOF. First, for u = 1 (i.e. v = 3), LSMTS(3) = $\{\{\langle a,b,0\rangle, \langle b,a,0\rangle\}\}$ is trivial. Next assume $u \geq 5$, notice $u \not\equiv 0 \pmod 2$, $u \not\equiv 0 \pmod 3$, it is easy to see that 2 and 3 are invertible in Z_u and $x+i \not\equiv -2x+i$ $(x \in Z_u\backslash\{0\})$.

<u>1</u>. Each of $\mathcal{B}_i$ is SMTS(v):

Simple counting shows that $\mathcal{B}_i$ contains $\frac{1}{3}(u+1)(u+2)$ cyclic triples, just the number expected. Therefore we only have to show that every ordered pair P of distinct elements of S is contained in some cyclic triple in $\mathcal{B}_i$. All the possibilities are exhausted as follows:

i) $P = (a,b), (b,a), (a,i), (i,a), (b,i), (i,b)$. It is easy to see that $P \subset \langle a,b,i\rangle$ or $\langle b,a,i\rangle$ in (1).

ii) $P = (a,k)$ $(k \in Z_u\backslash\{i\})$ is contained in $\langle a,x+i,-2x+i\rangle$ of (2), where $x = k-i \in Z_u\backslash\{0\}$; $P = (k,a)$ $(k \in Z_u\backslash\{i\})$ is contained in $\langle a,x'+i,-2x'+i\rangle$ of (2), where $x' = 2^{-1}(i-k) \in Z_u\backslash\{0\}$. It is similar for $P = (b,k)$, (k,b) $(k \in Z_u\backslash\{i\})$.

iii) $P = (k,j)$ $(k \neq j \in Z_u)$. If $k+j \equiv 3i-k$ or $3i-j$ (mod u), P is contained in $\langle a,k,j\rangle$ or $\langle b,k,j\rangle$ of (2) respectively; otherwise P is contained in $\langle k,j,s\rangle$ of (3), where $s \equiv 3i-j-k$ (mod u).

<u>2</u>. $\{\mathcal{B}_i;\ i \in Z_u\}$ is LSMTS(v):

We only have to show that every cyclic triple T of S is contained in some $\mathcal{B}_i$ above. All the possibilities are exhausted as follows:

i) $T = \langle a,b,k\rangle$, $\langle b,a,k\rangle$ $(k \in Z_u)$ is contained in (1) of $\mathcal{B}_k$.

ii) $T = \langle a,i,j\rangle$ $(i \neq j \in Z_u)$ is contained in (2) of $\mathcal{B}_k$, where $k \equiv 3^{-1}(2i+j)$ (mod u).

iii) $T = \langle b,i,j\rangle$ $(i \neq j \in Z_u)$ is contained in (2) of $\mathcal{B}_k$, where $k \equiv 3^{-1}(i+2j)$ (mod u).

iv) $T = \langle i,j,s\rangle$ $(i,j,s \in Z_u$ are pairwise distinct) is contained in (3) of $\mathcal{B}_k$, where $k \equiv 3^{-1}(i+j+s)$ (mod u).

This completes the proof. □

Theorem 1 proves the existence of all LMTS(v) of odd order v.

2. LSMTS(m+2) → LMTS(2m+2)

In this section we shall give a recursive construction for LMTS.

THEOREM 2. If there exists an LSMTS(m+2), then there exists an

LMTS(2m+2).

CONSTRUCTION. Let $\{(\{a,b\} \cup I_m, \mathcal{b}_i);\ i \in I_m\}$ be an LSMTS(m+2), where $I_m = \{0,1,\ldots,m-1\}$, $a,b \notin I_m$. Let $\langle a,b,i\rangle \in \mathcal{b}_i$ and set $S = \{a,b\} \cup (I_m \times I_2)$. We can construct 2m cyclic triple systems $\mathcal{B}_{ij}$ $(i \in I_m,\ j \in I_2)$ on S. Elements (x,0) and (x,1) of $I_m \times I_2$ are briefly denoted by x_0 and x_1 respectively.

$\mathcal{B}_{i,0}$ $(i \in I_m)$ consists of the following four parts:

(1) $\langle a,b,i_0\rangle$, $\langle b,a,i_1\rangle$, $\langle a,i_0,i_1\rangle$, $\langle b,i_1,i_0\rangle$;

(2) $\langle a,x_0,y_0\rangle$, $\langle a,x_1,y_1\rangle$, $\langle b,y_0,x_1\rangle$, $\langle b,y_1,x_0\rangle$ with $\langle a,x,y\rangle \in \mathcal{b}_i$, $x,y \in I_m$, this gives 4(m-1) cyclic triples;

(3) $\langle x_0,x_1,y_0\rangle$, $\langle x_1,x_0,y_1\rangle$ with $\langle a,x,y\rangle \in \mathcal{b}_i$, $x,y \in I_m$, this gives 2(m-1) cyclic triples;

(4) $\langle x_0,y_0,z_0\rangle$, $\langle x_0,y_1,z_1\rangle$, $\langle x_1,y_0,z_1\rangle$, $\langle x_1,y_1,z_0\rangle$ with $\langle x,y,z\rangle \in \mathcal{b}_i$, $x,y,z \in I_m$, this gives $\frac{4}{3}(m-1)(m-2)$ cyclic triples.

$\mathcal{B}_{i,1}$ $(i \in I_m)$ consists of the following four parts:

(1) $\langle a,b,i_1\rangle$, $\langle b,a,i_0\rangle$, $\langle a,i_1,i_0\rangle$, $\langle b,i_0,i_1\rangle$;

(2) $\langle a,x_0,y_1\rangle$, $\langle a,x_1,y_0\rangle$, $\langle b,y_0,x_0\rangle$, $\langle b,y_1,x_1\rangle$ with $\langle a,x,y\rangle \in \mathcal{b}_i$, $x,y \in I_m$;

(3) $\langle x_0,x_1,y_1\rangle$, $\langle x_1,x_0,y_0\rangle$ with $\langle a,x,y,\rangle \in \mathcal{b}_i$, $x,y \in I_m$;

(4) $\langle x_1,y_1,z_1\rangle$, $\langle x_1,y_0,z_0\rangle$, $\langle x_0,y_1,z_0\rangle$, $\langle x_0,y_0,z_1\rangle$ with $\langle x,y,z\rangle \in \mathcal{b}_i$, $x,y,z \in I_m$.

PROOF. First, since each $\mathcal{b}_i$ is a SMTS(m+2), therefore both $\langle a,b,i\rangle$ and $\langle b,a,i\rangle$ belong to $\mathcal{b}_i$, and conditions (2),(3) are actually $x,y \in I_m\backslash\{i\}$, both $\langle a,x,y\rangle$ and $\langle b,y,x\rangle$ belong to $\mathcal{b}_i$.

1. Each of $\mathcal{B}_{ij}$ is MTS(2m+2):

Direct calculation shows that $\mathcal{B}_{ij}$ contains (2m+1)(2m+2)/3 cyclic triples, just the number expected. Therefore we only have to show that every ordered pair P of S is contained in some cyclic triple in $\mathcal{B}_{ij}$. All the possibilities are exhausted as follows (we only give the proof for $\mathcal{B}_{i,0}$, the proof for $\mathcal{B}_{i,1}$ is similar):

i) P = (a,b),(b,a). It is easy to see that P is contained in (1).

ii) $P = (a,x_j),(x_j,b)$ $(x \in I_m,\ j \in I_2)$. If x = i, then P is contained in (1); otherwise there exists $y \in I_m\backslash\{i\}$ such that

$\langle a,x,y\rangle \in \mathcal{b}_i$, and P is contained in (2).

iii) $P = (b,x_j),(x_j,a)$ ($x \in I_m$, $j \in I_2$). If $x = i$, then P is contained in (1); otherwise there exists $y \in I_m\backslash\{i\}$ such that $\langle a,y,x\rangle \in \mathcal{b}_i$, and P is contained in (2).

iv) $P = (x_j,x_{1-j})$ ($x \in I_m$, $j \in I_2$). If $x = i$, then P is contained in (1); otherwise there exists $y \in I_m\backslash\{i\}$ such that $\langle a,x,y\rangle \in \mathcal{b}_i$, and P is contained in (3).

v) $P = (x_j,y_k)$ ($x \neq y \in I_m$, $j,k \in I_2$). There exists $z \in \{a,b\} \cup I_m$ such that $\langle x,y,z\rangle \in \mathcal{b}_i$. If $z = a$, P is contained in (2) (when $j = k$) or (3) (when $j \neq k$); if $z = b$, P is contained in (3) (when $j = k$) or (2) (when $j \neq k$); otherwise P is contained in (4).

2. $\{\mathcal{B}_{ij}: i \in I_m, j \in I_2\}$ is LMTS(2m+2):

We only have to show that every cyclic triple T of S is contained in some $\mathcal{B}_{ij}$ above. All the possibilities are exhausted as follows:

i) $T = \langle a,b,i_j\rangle$, $\langle a,i_j,i_{1-j}\rangle$ ($i \in I_m$, $j \in I_2$) is contained in (1) of $\mathcal{B}_{ij}$; $T = \langle b,a,i_j\rangle$, $\langle b,i_j,i_{1-j}\rangle$ ($i \in I_m$, $j \in I_2$) is contained in (1) of $\mathcal{B}_{i,1-j}$.

ii) $T = \langle a,x_j,y_k\rangle$ ($x \neq y \in I_m$, $j,k \in I_2$). There exists $i \in I_m$ such that $\langle a,x,y\rangle \in \mathcal{b}_i$. Let $s \equiv j+k \pmod 2$, then T is contained in (2) of $\mathcal{B}_{i,s}$. Similarly, $T = \langle b,x_j,y_k\rangle$ is contained in (2) of $\mathcal{B}_{i,1-s}$.

iii) $T = \langle x_j,x_{1-j},y_j\rangle$, $\langle x_j,x_{1-j},y_{1-j}\rangle$ ($x \neq y \in I_m$, $j \in I_2$). There exists $i \in I_m$ such that $\langle a,x,y\rangle \in \mathcal{b}_i$. Then T is contained in (3) of $\mathcal{B}_{i,0}$ or $\mathcal{B}_{i,1}$.

iv) $T = \langle x_j,y_j,z_j\rangle$, $\langle x_j,y_{1-j},z_{1-j}\rangle$ ($x,y,z \in I_m$ are pairwise distinct, $j \in I_2$). There exists $i \in I_m$ such that $\langle x,y,z\rangle \in \mathcal{b}_i$. Then T is contained in (4) of $\mathcal{B}_{ij}$.

This completes the proof. □

3. SOME LMTS(v) OF EVEN ORDER v

THEOREM 3. If $v \equiv 0$ or $4 \pmod{12}$, $v \geq 4$, then there exists an LMTS(v).

PROOF. Since $v+2 \equiv 2,6 \pmod{12}$, thus $(v+2)/2 \equiv 1,3 \pmod 6$ and $(v+2)/2 \geq 3$. By Theorem 1, we have an LSMTS((v+2)/2). By

Theorem 2 it follows that there exists an LMTS(v). □

LEMMA 1. If there exists an LMTS(v+1), then there exists an LMTS(3v+1) also.

LEMMA 2. If there exists an LMTS(v), then there exists an LMTS(3v) also.

For the proof of these two lemmas, see [2].

THEOREM 4. If $v \equiv 10\cdot 3^{k-1}$ or $34\cdot 3^{k-1} \pmod{4\cdot 3^{k+1}}$, then there exists an LMTS(v), where k is a positive integer.

PROOF. We use induction on k. First, by Theorem 3, if $u+1 \equiv 4,0 \pmod{12}$, then there exists an LMTS(u+1). And by Lemma 1, we have an LMTS(3u+1). Since $u \equiv 3,11 \pmod{12}$, $3u+1 \equiv 10,34 \pmod{36}$, therefore it is easy to see that Theorem 4 is true when $k = 1$.

Next, suppose for $w \equiv 10\cdot 3^{k-1}, 34\cdot 3^{k-1} \pmod{4\cdot 3^{k+1}}$ there exists an LMTS(w). By Lemma 2 it follows that when $v = 3w \equiv 10\cdot 3^{k}, 34\cdot 3^{k} \pmod{4\cdot 3^{k+2}}$ we have an LMTS(v). □

Up to now we have proved the existence of an LMTS(v) for $v \equiv 1,3 \pmod 6$ and for $v \equiv 0,4 \pmod{12}$. When $v \equiv 6,10 \pmod{12}$, we have also solved the case of $v \equiv 10,34 \pmod{36}$. In the next section we shall show that it only depends on the case of $v = 18$ and $v \equiv 6, 22 \pmod{36}$, $v > 6$ to complete the existence problem.

4. FURTHER IDEAS

LEMMA 3. Suppose for $v = 18$ and $v \equiv 6,22 \pmod{36}$, $v > 6$, there exists an LMTS(v). Then for $v \equiv 2\cdot 3^{k}, 22\cdot 3^{k-1} \pmod{4\cdot 3^{k+1}}$, $v > 6$, there exists an LMTS(v) also, where k is a positive integer.

The proof is similar to that of Theorem 4.

THEOREM 5. Suppose for $v = 18$ and $v \equiv 6,22 \pmod{36}$, $v > 6$, there exists an LMTS(v), then there exists an LMTS(v) for $v \not\equiv 2 \pmod 3$, $v > 1$, except $v = 6$.

PROOF.

1. If $v \equiv 1,3 \pmod 6$, $v > 1$, then by Theorem 1 it follows that there exists an LMTS(v).

2. If $v \equiv 0,4 \pmod{12}$, $v \geq 4$, then by Theorem 3 it follows that there exists an LMTS(v).

3. If $v \equiv 10 \pmod{12}$, when $v \equiv 10,34 \pmod{36}$ or $v \equiv 22 \pmod{36}$ we have an LMTS(v) by Theorem 4 or the supposition respectively.

4. If $v \equiv 6 \pmod{12}$, $v > 6$, let $v = 12t+6$ $(t \geq 1)$ and set $t = \sum_{i\geq 0} t_i 3^i$ $(t_i = 0,1,2)$. Let t_k $(k \geq 0)$ be the first number different from 1 in $t_0, t_1, t_2, \ldots$. Suppose δ is a nonnegative integer.

(1) If $t_k = 0$, then $t = \sum_{i=0}^{k-1} 3^i + 3^{k+1}\delta$, thus $v = 6(3^k-1)+4\cdot 3^{k+1}\delta+6 \equiv 2\cdot 3^{k+1} \pmod{4\cdot 3^{k+2}}$. By Lemma 3 there exists an LMTS(v).

2) If $t_k = 2$.

i) When $t_{k+1} = 0$, then $t = (3^k-1)/2+2\cdot 3^k+3^{k+2}\delta$, thus $v = 2\cdot 3^{k+1}+8\cdot 3^{k+1}+4\cdot 3^{k+3}\delta \equiv 10\cdot 3^{k+1} \pmod{4\cdot 3^{k+3}}$. By Theorem 4 there exists an LMTS(v).

ii) When $t_{k+1} = 1$, then $t = (3^k-1)/2+2\cdot 3^k+3^{k+1}+3^{k+2}\delta$, thus $v \equiv 22\cdot 3^{k+1} \pmod{4\cdot 3^{k+3}}$. By Lemma 3 there exists an LMTS(v).

iii) When $t_{k+1} = 2$, then $t = (3^k-1)/2+2\cdot 3^k+2\cdot 3^{k+1}+3^{k+2}\delta$, thus $v \equiv 34\cdot 3^{k+1} \pmod{4\cdot 3^{k+3}}$. By Theorem 4 there exists an LMTS(v).

This completes the proof. □

According to Theorem 5, the complete solution for large sets of disjoint Mendelsohn triple systems only depends on the orders of $v = 18$ and $v \equiv 6,22 \pmod{36}$, $v > 6$. To deal with these orders the next recursive construction in Theorem 6 will be of some help.

THEOREM 6. If there exists an LSMTS(p+2) and an LMTS(q+2), $q \geq 3$, then there exists an LMTS(pq+2).

CONSTRUCTION. Let $\{(\{a,b\} \cup I_p, \mathcal{B}_i): i \in I_p\}$ ($\langle a,b,i\rangle \in \mathcal{B}_i$) be an LSMTS(p+2) and $\{(\{a,b\} \cup Q, \mathcal{C}_j): j \in Q\}$ be an LMTS(q+2), where $Q = \{0,1,\ldots,q-1\}$ is an idempotent quasigroup of order q (its binary operation is denoted by "o"), $I_p = \{0,1,\ldots,p-1\}$, $a,b \notin I_p \cup Q$. Let $\alpha = (0,1,\ldots,q-1)$ be a cycle of order q. Now we can construct pq cyclic triple systems $\mathcal{T}_{ij}$ ($i \in I_p$, $j \in Q$) on the set $S = \{a,b\} \cup (I_p \times Q)$ as follows:

(1) $\langle (x,u),(y,v),(z,(u \circ v)\alpha^j)\rangle$ with $\langle x,y,z\rangle \in \mathcal{B}_i$, $x,y,z \in I_p$, $u,v \in Q$. This gives $q^2(p-1)(p-2)/3$ cyclic triples;

(2) $\langle (x,u),(x,v),(y,(u \circ v)\alpha^j)\rangle$ with $\langle a,x,y\rangle \in \mathcal{B}_i$, $x,y \in I_p$, $u \neq v \in Q$. This gives $(q^2-q)(p-1)$ cyclic triples;

(3) $\langle a,(x,u),(y,u\alpha^j)\rangle$, $\langle b,(y,u\alpha^j),(x,u)\rangle$ with $\langle a,x,y\rangle \in \mathcal{B}_i$, $x,y \in I_p$, $u \in Q$. This gives $2q(p-1)$ cyclic triples;

(4) $\langle (i,u),(i,v),(i,w)\rangle$ with $\langle u,v,w\rangle \in \mathcal{C}_j$ (whenever a or b appears for u,v,w, omit the first coordinate i). This gives $\frac{1}{3}(q+2)(q+1)$ cyclic triples.

PROOF. By the symmetry of LSMTS(p+2), it is easy to see that conditions (2),(3) are actually $x,y \in I_p \setminus \{i\}$, both $\langle a,x,y\rangle$ and $\langle b,y,x\rangle$ belong to $\mathcal{B}_i$.

1. Each $\mathcal{T}_{ij}$ is MTS(pq+2):

Direct calculation shows that each $\mathcal{T}_{ij}$ contains $\frac{1}{3}(pq+2)(pq+1)$ cyclic triples, just the number expected. Therefore we only have to show that every ordered pair P of distinct elements of S is contained in some cyclic triple in $\mathcal{T}_{ij}$. All the possibilities are exhausted as follows:

i) $P = (a,b)$. There exists $w \in Q$ such that $\langle a,b,w\rangle \in \mathcal{C}_j$, thus P is contained in $\langle a,b,(i,w)\rangle$ of (4). (It is similar for $P = (b,a)$).

ii) $P = (a,(x,u))$ ($x \in I_p$, $u \in Q$). If $x = i$, there exists $v \in \{b\} \cup Q \setminus \{u\}$ such that $\langle a,u,v\rangle \in \mathcal{C}_j$, when $v = b$, P is contained in $\langle b,a,(i,u)\rangle$ of (4); otherwise P in $\langle a,(i,u),(i,v)\rangle$ of (4). If $x \neq i$, there exists $y \in I_p$ such that $\langle a,x,y\rangle \in \mathcal{B}_i$, then P is contained in $\langle a,(x,u),(y,u\alpha^j)\rangle$ of (3). (It is similar for P = (b,(x,u)), ((x,u),a) and ((x,u),b)).

iii) P = ((x,u),(x,v)) ($x \in I_p$, $u \neq v \in Q$). If $x \neq i$, there exists $y' \in I_p$ such that $\langle a,x,y'\rangle \in \mathcal{B}_i$, then P is contained in $\langle (x,u),(x,v),(y',(u \circ v)\alpha^j)\rangle$ of (2). If $x = i$, there exists $w \in \{a,b\} \cup Q$ such that $\langle u,v,w\rangle \in \mathcal{b}_j$, then P is contained in (4).

iv) P = ((x,u),(y,v)) ($x \neq y \in I_p$, $u,v \in Q$). There exists $z \in \{a,b\} \cup I_p$ such that $\langle x,y,z\rangle \in \mathcal{B}_i$. If $z \in I_p$, then P is contained in $\langle (x,u),(y,v),(z,(u \circ v)\alpha^j)\rangle$ of (1). If z = a, then when $v\alpha^{-j} = u$, P is contained in $\langle a,(x,u),(y,u\alpha^j)\rangle$ of (3); otherwise there exists $v' \neq u$ such that $v' \circ u = v\alpha^{-j}$, P is contained in $\langle (x,v'),(x,u),(y,(v' \circ u)\alpha^j)\rangle$ of (2). If z = b, then when $u\alpha^{-j} = v$, P is contained in $\langle b,(x,v\alpha^j),(y,v)\rangle$ of (3); otherwise there exists $u' \neq v$ such that $v \circ u' = u\alpha^{-j}$, P is contained in $\langle (y,v),(y,u'),(x,(v \circ u')\alpha^j)\rangle$ of (2).

<u>2</u>. $\{\mathcal{T}_{ij}: i \in I_p, j \in Q\}$ is LMTS(pq+2):

We only have to show that every cyclic triple T of S is contained in some $\mathcal{T}_{ij}$. All the possibilities are exhausted as follows:

i) $T = \langle a,b,(i,w)\rangle$ ($i \in I_p$, $w \in Q$). There exists $j \in Q$ such that $\langle a,b,w\rangle \in \mathcal{b}_j$, then T is contained in (4) of $\mathcal{T}_{ij}$. (It is similar for $T = \langle b,a,(i,w)\rangle$).

ii) $T = \langle a,(i,v),(i,w)\rangle$ ($i \in I_p$, $v \neq w \in Q$). There exists $j \in Q$ such that $\langle a,v,w\rangle \in \mathcal{b}_j$, then T is contained in (4) of $\mathcal{T}_{ij}$. (It is similar for $T = \langle b,(i,v),(i,w)\rangle$).

iii) $T = \langle a,(x,u),(y,v)\rangle$ ($x \neq y \in I_p$, $u,v \in Q$). There exists $i \in I_p$, $j \in Q$ such that $\langle a,x,y\rangle \in \mathcal{B}_i$, $u = v\alpha^j$, then T is contained in (3) of $\mathcal{T}_{ij}$. (It is similar for $T = \langle b,(x,u),(y,v)\rangle$).

iv) $T = \langle (i,u),(i,v),(i,w)\rangle$ ($i \in I_p$, $u,v,w \in Q$ are pairwise distinct). There exists $j \in Q$ such that $\langle u,v,w\rangle \in b_j$, then T is contained in (4) of $\mathcal{T}_{ij}$.

v) $T = \langle (x,u),(x,v),(y,w)\rangle$ ($x \neq y \in I_p$, $u,v,w \in Q$, $u \neq v$). There exists $i \in I_p$, $j \in Q$ such that $\langle a,x,y\rangle \in \mathcal{B}_i$, $(u\circ v)\alpha^j = w$, then T is contained in (2) of $\mathcal{T}_{ij}$.

vi) $T = \langle (x,u),(y,v),(z,w)\rangle$ ($x,y,z \in I_p$ are pairwise distinct, $u,v,w \in Q$). There exists $i \in I_p$, $j \in Q$ such that $\langle x,y,z\rangle \in \mathcal{B}_i$, $(u\circ v)\alpha^j = w$, then T is contained in (1) of $\mathcal{T}_{ij}$.

This completes the proof. □

Finally we would like to point out that if $p \equiv \pm 1 \pmod 6$ and there exists an LMTS(q+2) for $q \equiv \pm 4 \pmod{12}$, then Theorem 6 gives some new results. For instance, there exist LMTS(48k+10), LMTS(48k+42), etc.

ACKNOWLEDGEMENTS

We thank Prof. Wu Li-sheng for drawing our attention to the problem. We also wish to thank Prof. Zhu Lie for some helpful comments regarding the presentation.

REFERENCES

1. N.S. Mendelsohn, A natural generalization of Steiner triple systems, in "Computers in Number Theory," (Academic Press, New York, 1971), 323-338.

2. C.C. Lindner, On the number of disjoint Mendelsohn triple systems, J. Combinatorial Theory, 30A(1981), 320-330.

3. Lu Jia-xi, On large sets of disjoint Steiner triple systems VI, J. Combinatorial Theory, 37 A(1984), 198-192.

4. C.J. Colbourn and M.J. Colbourn, Disjoint cyclic Mendelsohn triple systems, Ars Combinatoria 11(1981), 3-8.

5. Wu Li-sheng, On large sets of disjoint Mendelsohn triple systems, J. Suzhou University 3(1987), 111-114.

Simple BIB Designs and 3-Designs of Small Orders

KU Tung-Hsin, The Hefei Branch Research Group of Combinatorial Mathematics Academia Sinica, Hefei, Anhui, China

1. INTRODUCTION

A $t\text{-}(v,k;\lambda)$ design is an ordered pair (X,β) where X is a finite set containing v elements and β is a collection of k-subsets (called blocks) of X such that each t-subset of X is contained in exactly λ blocks; a $t\text{-}(v,k;\lambda)$ design is called simple if it contains no repeated blocks.

For $0 \leq i < t$, let

$$\lambda_i = \frac{\lambda\binom{v-i}{t-i}}{\binom{k-i}{t-i}} = \lambda\frac{(v-i)(v-i-1)\ \ldots\ (v-t+1)}{(k-i)(k-i-1)\ \ldots\ (k-t+1)} .$$

It is well known that a necessary condition for the existence of a $t\text{-}(v,k;\lambda)$ design is that the numbers λ_i must be integers for all $0 \leq i < t$. It is easy to see that the following is an additional condition for the existence of a simple $t\text{-}(v,k;\lambda)$ design:

$$\lambda \leq \frac{\binom{k}{t} \cdot \binom{v}{k}}{\binom{v}{t}} .$$

Thus, in the case t = 2, the following are necessary conditions for the existence of a simple 2-(v,k;λ) design:

$$\begin{aligned} \lambda(v-1) &\equiv 0 \quad (\text{mod } (k-1)) \\ \lambda v(v-1) &\equiv 0 \quad (\text{mod } k(k-1)) \qquad (1) \\ \lambda &\leq \binom{v-2}{k-2} \end{aligned}$$

In the case t = 3, the following are necessary conditions for the existence of a simple 3-(v,k;λ) design:

$$\begin{aligned} \lambda v(v-1)(v-2) &\equiv 0 \quad (\text{mod } k(k-1)(k-2)) \\ \lambda(v-1)(v-2) &\equiv 0 \quad (\text{mod } (k-1)(k-2)) \qquad (2) \\ \lambda(v-2) &\equiv 0 \quad (\text{mod } (k-2)) \\ \lambda &\leq \binom{v-3}{k-3} \end{aligned}$$

H.-D.O.F. Gronau ([1]) listed the known results on the existence of simple t-(v,k;λ) designs for $6 \leq v \leq 16$. H. Shen ([2]) constructed all the simple 2-(v,k;λ) designs for the smallest possible λ for which the existence of a simple 2-(v,k;λ) design is previously unknown.

It is the purpose of the present paper to construct a few more new simple 2-(v,k;λ) designs and 3-(v,k;λ) designs of small orders.

2. SIMPLE BIB DESIGNS OF SMALL ORDERS

Let (Z_v, β) be a t-(v,k;λ) design. For $B = \{a_1, a_2, \ldots, a_k\} \in \beta$, let

$$B+x = \{a_1+x, a_2+x, \ldots, a_k+x\}, \qquad x \in Z_v.$$

If for each $B \in \beta$ and each $x \in Z_\nu$, we have $B+x \in \beta$, then (Z_v, β) is called a cyclic t-(v,k;λ) design.

Instead of constructing cyclic simple t-(v,k;λ) designs on Z_v, in some cases, we can construct simple 3-(v,k;λ) designs on the set $Z_{v-1} \cup \{\infty\}$, where ∞ is an element which is fixed under the action of the cyclic additive group Z_{v-1}. In this case, we let

$$x+\infty = \infty, \quad \text{for all} \quad x \in Z_{v-1}.$$

The designs obtained in this way are called *rotational.*

In this section we will construct some cyclic or rotational simple BIB designs of small orders which are previously unknown.

For given v and k, let λ_0 be the smallest possible positive integer such that there exists a simple t-$(v,k;\lambda_0)$ design. Let $s = \binom{v-t}{k-t}/\lambda_0$, if there exist s pairwise disjoint simple t-$(v,k;\lambda_0)$ designs, then we call such a set of s pairwise disjoint simple t-$(v,k;\lambda_0)$ designs a *large set* of disjoint simple t-$(v,k;\lambda_0)$ designs.

Let v = 11. By (1), the necessary condition for the existence of simple 2-$(11,4;\lambda)$ designs is

$$\lambda = 6s, \quad 1 \leq s \leq 6. \tag{3}$$

We will construct a simple 2-$(11,4;\lambda)$ design for each such λ.

THEOREM 1. There exists a large set of disjoint simple 2-(11,4;6) designs.

PROOF. We form six pairwise disjoint cyclic 2-(11,4;6) designs without repeated blocks by developing the following sets mod 11:

(I)	{0,1,2,4},	{0,1,2,6},	{0,1,3,6},	{0,1,5,8},	{0,2,4,8}
(II)	{0,1,2,7},	{0,1,2,9},	{0,1,4,7},	{0,1,6,9},	{0,2,4,7}
(III)	{0,1,2,5},	{0,1,3,5},	{0,1,3,7},	{0,1,3,8},	{0,1,4,6}
(IV)	{0,1,2,8},	{0,1,4,9},	{0,1,5,9},	{0,1,6,8},	{0,1,7,9}
(V)	{0,1,2,3},	{0,1,4,8},	{0,1,5,6},	{0,2,4,6},	{0,2,5,8}
(VI)	{0,1,3,4},	{0,1,3,9},	{0,1,4,5},	{0,1,5,7},	{0,2,5,7}

These six 2-(11,4;6) designs form a large set.

As a consequence, we have proved that the necessary condition for the existence of a simple 2-$(11,4;\lambda)$ design is also sufficient.

COROLLARY. There exists a simple 2-$(11,4;\lambda)$ design if and only if

$$\lambda \equiv 0 \pmod{6}, \qquad 6 \le \lambda \le 36. \tag{4}$$

Now, let us consider the existence of simple $2\text{-}(14,4;\lambda)$ designs.

THEOREM 2. There exists a simple $2\text{-}(14,4;\lambda)$ design if and only if

$$\lambda = 6s, \qquad 6 \le \lambda \le 66. \tag{5}$$

PROOF. We form seven disjoint simple rotational $2\text{-}(14,4;6)$ designs on $Z_{13} \cup \{\infty\}$ as follows:

(i) {0,1,2,10}, {0,1,3,9}, {0,1,3,10}, {0,1,5,8}, {0,1,4,6}, {0,2,6,∞}, {0,2,7,∞} (mod 13)

(ii) {0,1,2,3}, {0,1,4,7}, {0,1,6,9}, {0,2,4,8}, {0,2,5,8}, {0,1,9,∞}, {0,2,9,∞} (mod 13)

(iii) {0,1,3,8}, {0,1,3,11}, {0,1,5,9}, {0,1,7,10}, {0,2,6,8}, {0,1,2,∞}, {0,3,7,∞} (mod 13)

(iv) {0,1,2,4}, {0,1,3,6}, {0,1,5,10}, {0,1,6,8}, {0,2,6,9}, {0,1,4,∞}, {0,2,8,∞} (mod 13)

(v) {0,1,2,5}, {0,1,3,5}, {0,1,6,7}, {0,2,4,9}, {0,2,7,10}, {0,1,10,∞}, {0,3,9,∞} (mod 13)

(vi) {0,1,2,6}, {0,1,3,7}, {0,2,5,9}, {0,2,5,10}, {0,1,7,11}, {0,1,3,∞}, {0,1,5,∞} (mod 13)

(vii) {0,1,4,10}, {0,1,4,11}, {0,1,2,8}, {0,1,9,11}, {0,2,5,7}, {0,4,8,∞}, {0,1,8,∞} (mod 13)

Using s of these pairwise disjoint simple $2\text{-}(14,4;6)$ designs we can obtain simple $2\text{-}(14,4;6s)$ designs for $s < 7$.

As all the 4-subsets of $X = Z_{13} \cup \{\infty\}$ form a simple $2\text{-}(14,4;66)$ design, from any simple $2\text{-}(14,4;\lambda)$ design we can obtain a simple $2\text{-}(14,4;66-\lambda)$.

3. CYCLIC OR ROTATIONAL 3-DESIGNS WITHOUT REPEATED BLOCKS

In this section, we will construct cyclic simple $3\text{-}(11,4;\lambda)$ designs for $\lambda = 4$ and 8 and construct a cyclic simple $3\text{-}(13,4;2)$ design and a rotational $3\text{-}(12,4;\lambda)$ design for $\lambda = 3,6$ or 9.

The smallest positive integers v and k for which the existence of a simple 3-$(v,k;\lambda)$ design remains unknown are v = 11 and k = 4. By the necessary condition (2), if a simple 3-$(11,4;\lambda)$ design exists, then λ = 4 or 8. We will construct two cyclic simple 3-(11,4;4) designs (X,β_1) and (X,β_2), on the set Z_{11} as follows:

(I) $X = Z_{11}$

β_1: {0,1,2,3},{0,1,4,8},{0,1,5,6},{0,2,4,6},{0,2,5,8}
{0,1,2,8},{0,1,3,7},{0,1,4,6},{0,1,3,5},{0,1,3,8} (mod 11)
{0,1,2,5},{0,1,5,9},{0,1,6,8},{0,1,7,9},{0,1,4,9}

(II) $X = Z_{11}$

β_2: {0,1,2,4},{0,1,2,6},{0,1,2,7},{0,1,2,9},{0,1,3,4}
{0,1,3,6},{0,1,3,9},{0,1,4,5},{0,1,4,7},{0,1,5,7} (mod 11)
{0,1,5,8},{0,1,6,9},{0,2,4,7},{0,2,4,8},{0,2,5,7}

It can be checked that β_1 and β_2 have no common blocks. Hence $(X,\beta_1 \cup \beta_2)$ is a cyclic simple 3-(11,4,8) design. Thus we have proved the following theorem:

THEOREM 3. There exists a simple 3-$(11,4;\lambda)$ design if and only if λ = 4 or 8.

By the necessary condition (2), we can construct cyclic simple 3-(13,4;2) design (X,β) on the set Z_{13} as follows:

$X = Z_{13}$

β: {0,1,2,3}, {0,1,3,4}, {0,1,4,5}, {0,1,5,6},
{0,1,6,7}, {0,2,4,8}, {0,2,4,9}, {0,2,5,7}, (mod 13)
{0,2,6,9}, {0,2,5,10}, {0,3,6,9}

It is obvious that all the 4-subsets of Z_{13} form a simple 3-(13,4;10) design. Thus, we have proved the following theorem:

THEOREM 4. There exists a simple 3-$(13,4;\lambda)$ design, for λ = 2, 8 or 10.

Now, we will determine the existence of a simple 3-$(12,4;\lambda)$ design.

If there exists a simple 3-$(12,4;\lambda)$ design, then by the necessary condition (2), we have:

$$\lambda \equiv 0 \pmod 3), \qquad \lambda \leq 9. \tag{6}$$

Let $X = Z_{11} \cup \{\infty\}$. We construct a rotational simple 3-(12,4;3) design (X,β_1) and a rotational simple 3-(12,4;6) design (X,β_2) as follows:

(I) A rotational simple 3-(12,4;3) design:

$X = Z_{11} \cup \{\infty\}$

β_1: {0,1,2,3}, {0,1,2,4}, {0,1,3,7}, {0,1,4,5}, {0,1,4,8}, {0,1,5,6}, {0,1,6,9}, {0,2,4,6}, {0,2,5,7}, {0,2,5,8}, {0,1,5,∞}, {0,1,8,∞}, {0,1,9,∞}, {0,2,5,∞}, {0,2,7,∞} (mod 11)

(II) A rotational simple 3-(12,4;6) design:

$X = Z_{11} \cup \{\infty\}$

β_2: {0,1,2,5}, {0,1,2,6}, {0,1,2,7}, {0,1,2,8}, {0,1,2,9}, {0,1,3,4}, {0,1,3,5}, {0,1,3,6}, {0,1,3,8}, {0,1,3,9}, {0,1,4,6}, {0,1,4,7}, {0,1,4,9}, {0,1,5,7}, {0,1,5,8}, {0,1,5,9}, {0,1,6,8}, {0,1,7,9}, {0,2,4,7}, {0,2,4,8} {0,1,2,∞}, {0,1,3,∞}, {0,1,4,∞}, {0,1,6,∞}, {0,1,7,∞}, {0,2,4,∞}, {0,2,6,∞}, {0,2,8,∞}, {0,3,6,∞}, {0,3,7,∞} (mod 11)

It can be checked that $\beta_1 \cap \beta_2 = \Phi$, so $(X,\beta_1 \cup \beta_2)$ is a rotational simple 3-(12,4;9) design. Thus, we have proved the following theorem:

THEOREM 5. There exists a simple 3-(12,4;λ) design if and only if λ = 3,6 or 9.

REFERENCES

[1] H.D.O.F. Gronau, A survey of results on the numbers of t-(v,k,λ) designs, Ann. Discrete Math. 26(1985), 209-219.

[2] H. Shen, Indecomposable simple 2-(v,k,λ) designs of small orders, J. Comb. Math. Comb. Comp. (to appear).

[3] H. Hanani, Balanced incomplete block designs and related designs, Discrete Math. 11(1975), 255-369.

Characterizing Strongly Chordal Graphs by Using Minimal Relative Separators

Shaohan MA, Department of Computer Sciences, Shandong University, Jinan, People's Republic of China

Julin WU, Department of Mathematics, Qingdao University, Qingdao, People's Republic of China

1. INTRODUCTION

Notations and terminologies about graphs are from [2] if they are not defined here, and those about hypergraphs are from [1]. A graph is represented as $G = (V,E)$, where V is the vertex set and E is the edge set. A graph G is *chordal* if each of its cycles of length greater than or equal to 4 has a chord. Equivalently, G is chordal if no cycle C_n $(n \geq 4)$ is an induced subgraph of G. In this paper, subgraphs always mean vertex-induced subgraphs. The subgraph induced by the vertex subset S is denoted by $G[S]$. A *clique* is a complete subgraph, which is usually denoted by its vertex set. The set $N_G(v)$ is the set of adjacent vertices of a vertex v in G; $N_G[v] = N_G(v) \cup \{v\}$ is called the *closed neighborhood* of v. We often omit the index G if there is no confusion.

A *path* P in a graph G is a sequence of distinct vertices $(v_0,v_1,\ldots,v_p)$ such that $[v_i,v_{i+1}]$ is an edge of G for $i = 0,1,$

...,p-1. We say that this path *connects* x to y, or P is an x-y path. p is the length of the path P. The path P is *chordless* if P is the subgraph induced by $\{v_0, v_1, \ldots, v_p\}$. For two vertices x and y, the minimum number of edges of an x-y path is called the *distance* from x to y, and is denoted by $d_G(x,y)$.

Chordal graphs are studied in [3-8]. Many characterizations are known. One of them is that a graph G is chordal if and only if each minimal relative separator of G is a clique. Classes of graphs such as threshold graphs, interval graphs, trees and strongly chordal graphs are subclasses of chordal graphs. A graph G is *strongly chordal* if G is chordal and each even cycle of G admits a strong chord. A strong chord of an even cycle is a chord that splits the cycle into two paths of odd length.

In this paper, we characterize strongly chordal graphs by minimal relative separators. We show that a graph G is strongly chordal if and only if G is chordal and the minimal relative separator hypergraph of G is totally balanced. An algorithm to find all the minimal relative separators of a chordal graph is also given.

2. Characterization

Let a and b be two non-adjacent vertices in a connected component of G, and S be a vertex subset of G not containing a or b. S is called an a-b *relative separator* if a and b are not connected in G-S. S is an a-b minimal relative separator if S is an a-b relative separator but any proper subset of S is not. A vertex subset S of G is a minimal relative separator of G if S is an a-b minimal relative separator for some a and b. Rose [6] proved the following

THEOREM A. A graph G is chordal if and only if each minimal relative separator of G is a clique.

A *trampoline* of order p, T_p, is a graph on 2p vertices, $\{x_1, x_2, \ldots, x_p, y_1, y_2, \ldots, y_p\}$. The set $\{y_1, y_2, \ldots, y_p\}$ is a clique and x_i is only adjacent to y_i and y_{i+1}, where $y_{p+1} = y_1$.

Farber [4] gave the following

THEOREM B. A graph G is strongly chordal if and only if it is chordal and it contains no trampolines as subgraphs.

Let $\mathcal{S} = \mathcal{S}(G)$ be the set of minimal relative separators of G. The hypergraph $H(G) = (V(G), \mathcal{S}(G))$ is called the *minimal relative separator hypergraph* of G, or MRS-hypergraph of G.

A cycle $C = (e_1, E_1, e_2, E_2, \ldots, e_p, E_p)$ in a hypergraph is a *special cycle* if each edge E_i of C contains exactly two vertices of C. A hypergraph is *totally balanced* if it contains no special cycle of length ≥ 3. One of the main results of this paper is

THEOREM 1. A chordal graph G is strongly chordal if and only if the MRS-hypergraph of G is totally balanced.

We need some preliminaries to prove the theorem.

DEFINITION. Two vertices x and y are called *semi-adjacent* if the length of each chordless x-y path is 2.

LEMMA. Let S be a vertex subset of a chordal graph G. Then S is a minimal relative separator of G if and only if $S = N(x) \cap N(y)$ for a pair of semi-adjacent vertices x and y of G.

PROOF. The sufficiency is obvious since S is an x-y minimal relative separator if the condition holds. Now we prove the necessity. Let S be an a-b minimal relative separator in G, and G_a, G_b be the connected components of G-S containing respectively a and b. We will show that $S = N(x) \cap N(y)$ for some $x \in G_a$ and $y \in G_b$.

In fact, let x be a vertex in G_a with maximum number of neighbors in S. Then $S \subseteq N(x)$. For, assume that it was not true; let $s \in S-N(x)$. There is a chordless s-x path P whose inner vertices are all in G_a. Let x′ be the neighbor of s on P. Then x′ is adjacent to each vertex of $N(x) \cap S$ (otherwise there would be a chordless cycle of length ≥ 4). This is contrary to the choice of x. Similarly, if y is a vertex in G_b with maximum number of neighbors in S, then $S \subseteq N(y)$. It is easy to see that x and y are semi-adjacent and $S = N(x) \cap N(y)$. □

PROOF OF THEOREM 1. Sufficiency: If G is not strongly chordal, then G contains a trampoline T_p. Without loss of generality, let T_p be a trampoline with minimum order. $V(T_p) = \{x_1, x_2, \ldots, x_p, y_1, y_2, \ldots, y_p\} = X \cup Y$. An example is shown in Figure 1.

Choose an x_1-y_p minimal relative separator S_1. Then $y_1, y_2 \in S_1$. Assume that S_1 contains other y_i's of Y other than y_1, y_2, y_p. There is a chordless x_1-y_i path P none of whose inner vertices is in S_1. Let z be the neighbor of y_i on P. Then z is not adjacent to y_p. z is adjacent to y_1, y_2 since G is chordal.

Let y_j be the vertex of Y adjacent to z with maximum index j. Then $p > j \geq 3$ and $\{y_1, y_j, y_{j+1}, \ldots, y_p, z, x_j, x_{j+1}, \ldots, x_p\}$ induces a trampoline of order p-j+2 in G. This is contrary to the choice of T_p.

So the intersection of Y with any x_1-y_p minimal relative separator S_1 is $\{y_1, y_2\}$. Similarly, the intersection of Y with any x_i-y_{i-1} minimal relative separator S_i is $\{y_i, y_{i+1}\}$. $(y_1, S_1, y_2, S_2, \ldots, y_p, S_p)$ is a special cycle of length p in H(G). The MRS-hypergraph of G is not totally balanced.

Necessity: If $H(G) = (V(G), \mathcal{S}(G))$ is not totally balanced, there must be a special cycle $C = (y_1 S_1, y_2, S_2, \ldots, y_p, S_p)$. Without loss of generality, we assume that the special cycle is of minimum length.

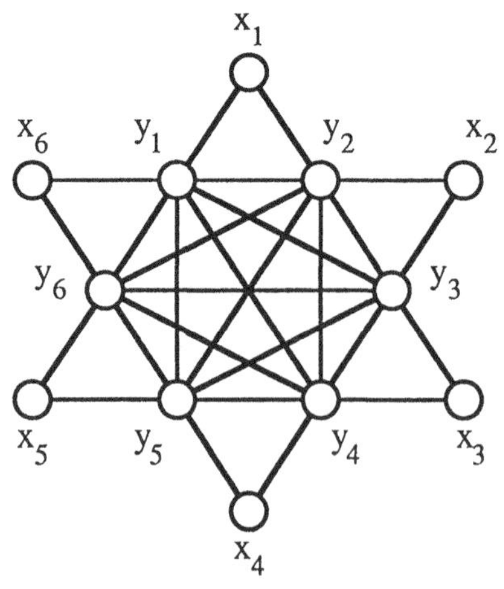

Figure 1. A trampoline of order 6

Let $Y = \{y_1,y_2,\ldots,y_p\}$. Y is a clique of G: otherwise, if y_i and y_j are not adjacent and S is a y_iy_j minimal relative separator, then we can find a smaller special cycle in $H(G) = (V(G),\mathcal{S}(G))$ according to C and S.

By the lemma we know that there is a pair of semi-adjacent vertices x_1' and x_1'' such that $S_1 = N(x_1') \cap N(x_1'')$. x_1' and x_1'' are adjacent to y_1 and y_2. If x_1' is adjacent to other vertices y_i's of Y, x_1' must be adjacent to all the vertices of Y. In fact, if x_1' is not adjacent to y_j and S is an x_1'-y_j minimal relative separator, then we can find a smaller special cycle in $H(G) = (V(G),\mathcal{S}(G))$ according to C and S. When x_1' is adjacent to all the vertices of Y, $Y \cap N(x_1'') = \{y_1,y_2\}$.

Therefore there is a vertex x_1 whose only neighbors in Y are y_1 and y_2. Similarly, there is a vertex x_i whose only neighbors in Y are y_i and y_{i+1}. The subgraph of G induced by $\{x_1,x_2,\ldots,x_p,y_1,y_2,\ldots,y_p\}$ is T_p. G is not strongly chordal. □

3. FINDING ALL THE MINIMAL RELATIVE SEPARATORS

Let n be the number of vertices of a graph G. An ordering of G is a bijection σ from V to $\{1,2,\ldots,n\}$. $\sigma(v)$ is called the *number* of v. For simplicity, when considering a specified ordering, we use v_i to represent the vertex with number i, and G_i, $MN(v_i)$, $MN[v_i]$ mean the following respectively:

$$G_i = G[v_i,v_{i+1},\ldots,v_n];$$
$$MN(v_i) = \{v_j \in N[v_i]: j > i\};$$
$$MN[v_i] = \{v_j \in N[v_i]: j \geq i\}.$$

An ordering of G is called a *perfect elimination ordering* if each $MN[v_i]$ is a clique.

Rose [6] proved that a graph G is chordal if and only if there is a perfect elimination ordering of G. Rose, Tarjan and Leuker [7] gave an algorithm, ALGORITHM Lex BFS, to test the chordality of a graph. If the graph is chordal, the algorithm gives a perfect elimination ordering. We now present this algorithm. Here label(v) is a vector attached to vertex v.

```
        ALGORITHM Lex BFS
Input: A chordal graph G = (V,E)
begin
1       assign the label φ to each vertex of G;
2       for i = n,n-1,...,2, 1 do
        begin
3           select: pick an unnumbered vertex v with
                lexicographically largest label;
4           σ(v):=i; comment: This assigns to v the number i.
            m(v):= "the smallest number in label(v)";
            D(v):= "the number of numbers in label(v)"
5           update: for each unnumbered vertex w ∈ N(v),
                add i to the end of label(v)
        end
end
```

When the algorithm stops, σ is a perfect elimination ordering, D(i) is the number of vertices in $MN(v_i)$ and m(i) is the smallest number in $MN(v_i)$.

Let v be a vertex of graph G. v is *simplicial* if N(v) is a clique. v is *subsimplicial* if v is simplicial but N(v) is not a maximal clique. By this definition we know that σ is a perfect elimination ordering if and only if v_i is a simplicial vertex of G_i. For the ordering produced by ALGORITHM Lex BFS, a slightly stronger result holds. In [8] it was shown that for the ordering σ produced by ALGORITHM Lex BFS, v_i is a subsimplicial vertex for $1 \leq i \leq n$.

Let u,v be two vertices of a graph G. If N[u] = N[v], u is called a *twin* of v. A property of the ordering σ produced by ALGORITHM Lex BFS was pointed out in [8]:

THEOREM C. Let σ be the ordering produced by ALGORITHM Lex BFS for a chordal graph G. If v_j is a twin of v_i in G_i, then any v_k $(i < k < j)$ must be a twin of v_i in G_i.

Now we give the algorithm to find all the minimal relative separators of a chordal graph.

ALGORITHM MRS

Input: A chordal graph $G = (V,E)$ and the σ, D and M produced in ALGORITHM Lex BFS

Begin

1 for i = 1,2,...,n do

begin

2 If $D(i) = n-i$ then stop; comment: G_i is complete.

3 If $m(i) \neq i+1$ or $D(i) \neq D(i+1)+1$ then output $MN(v_i)$

end

end

THEOREM 2. The output of ALGORITHM MRS is exactly the set of minimal relative separators of G.

PROOF. It suffices to notice the following three lemmas.

LEMMA 1. If v is simplicial in G and u is a twin of v, then G and G-u have the same set of minimal relative separators.

LEMMA 2. The condition in the third row of the algorithm holds if and only if v_i has no twin in G.

LEMMA 3. If v is a subsimplicial vertex of G and there are no twins of v in G, then N(v) is a minimal relative separator of G and $\mathcal{S}(G) = \{N(v)\} \cup \mathcal{S}(G-v)$.

The first Lemma is obvious. Now we prove the other two.

PROOF OF LEMMA 2. If there is a twin of v_i in G_i, v_{i+1} must be a twin of v_i in G_i by Theorem C. So v_i is adjacent to v_{i+1}, $m(i) = i+1$ and $D(i) = D(i+1)+1$.

On the other hand, if $m(i) = i+1$ and $D(i) = D(i+1)+1$, then v_i is adjacent to v_{i+1}, v_i and v_{i+1} have the same number of neighbors in G_i. v_{i+1} must be a twin of v_i otherwise there would be a vertex a in G_i that is adjacent to v_i while a is not adjacent to v_{i+1}. This is contrary to the fact that v is simplicial in G_i. □

PROOF OF LEMMA 3. In the proof of Lemma 3, v is always a subsimplicial vertex and v has no twins in G.

We know by the lemma of Theorem 1 that the set $\mathcal{S}$ of minimal relative separators of G is $\{N(x) \cap N(y)$: x and y are semi-adjacent$\}$. Now we consider the semi-adjacent vertices of G.

Since N(v) is not a maximal clique in G-v, there must be a vertex $u \in V(G)-N[v]$ such that $N(v) \subseteq N(u)$. It is easy to see by the definition that N(v) is a u-v minimal relative separator.

Let y be a semi-adjacent vertex of v in G. $y \neq u$. If $N(v) \subset N(y)$ then $N(v) \cap N(y) = N(v)$. If $N(v) \not\subset N(y)$ then $N(y) \subset N(v)$ (otherwise there is a v-y chordless path of length 3). u and y are semi-adjacent in G-v and $N(y) \cap N(v) = N_{G-v}(y) \cap N_{G-v}(u)$. That is, $N(y) \cap N(v)$ is also a minimal relative separator of G-v.

Let x and y be semi-adjacent, $x,y \neq u,v$. Then x and y are semi-adjacent in G-v and $N(x) \cap N(y) = N_{G-v}(X) \cap N_{G-v}(y)$. That is, $N(x) \cap N(y)$ is also a minimal relative separator of G-v.

It is easy to see that a pair of semi-adjacent vertices of G-v is also a pair of semi-adjacent vertices of G and $N_{G-v}(x) \cap N_{G-v}(y) = N_G(x) \cap N_G(y)$. So we have proved the third lemma. □

This completes the proof of Theorem 2.

COROLLARY. The number of minimal relative separators of a chordal graph is less than or equal to the number of vertices of G.

REFERENCES

[1] C. Berge, Graphs and Hypergraphs, North-Holland, Amsterdam. 1973.

[2] J.A. Bondy and U.S.R. Murty, Graph Theory with Applications, Macmillan Press Ltd. 1976.

[3] P. Buneman, A characterization of rigid-circuit graphs, Discrete Math. 9(1974), 205-212.

[4] M. Farber, Characterizations of strongly chordal graphs, Discrete Math. 43(1983), 173-189.

[5] F. Gavril, The intersection graphs of subtrees in trees are exactly chordal graphs, J. Combinatorial Theory 16B(1974), 47-56.

[6] D.J. Rose, Triangulated graphs and the elimination process, J. Math. Anal. Appl. 32(1970), 597-609.

[7] D.J. Rose, R.E. Tarjan and G.S. Leuker, Algorithmic aspects of vertex elimination on graphs, SIAM J. Comp. 5(1976), 266-283.

[8] J. Wu, Maximum strongly independent sets in chordal graphs, (submitted)

Constructions and Uses of Labeled Resolvable Designs

SHEN Hao, Department of Applied Mathematics, Jiao Tong University, Shanghai, P.R. China

1. INTRODUCTION

A balanced incomplete block design $S_\lambda(2,k,v)$ is an ordered pair $(X,\mathbb{B})$ where X is a finite set containing v elements, and $\mathbb{B}$ is a collection of k-subsets (called blocks) of X such that each pair of distinct elements of X is contained in exactly λ blocks.

A parallel class in an $S_\lambda(2,k,v)$ is a set of blocks which partitions X. An $S_\lambda(2,k,v)$ is called resolvable and denoted $RS_\lambda(2,k,v)$ if the blocks can be partitioned into parallel classes.

An almost parallel class in an $S_\lambda(2,k,v)$ is a set of blocks which partitions $X\backslash\{x\}$ for some $x \in X$. An $S_\lambda(2,k,v)$ is called almost resolvable and denoted $ARS_\lambda(2,k,v)$ if the blocks can be partitioned into almost parallel classes.

In this paper, we will introduce a new kind of resolvable (or almost resolvable) block designs -- labeled resolvable (or labeled almost resolvable) block designs, which will prove useful in the construction of resolvable block designs and resolvable group divisible designs.

Let $\lambda = m$ and let $(X,\mathbb{B})$ be an $RS_m(2,k,v)$. Without loss of generality, we may let $X = Z_v$ and for each block B,

$$B = \{a_1,a_2,\ldots,a_k\}.$$

Let a_i be the least non-negative representative of the residue class containing it. Then

$$0 \le a_1 < a_2 < \cdots < a_k \le v-1.$$

Let

$$\varphi\colon \mathbb{B} \to Z_m^{\binom{k}{2}}$$

be a mapping where for each $B = \{a_1,a_2,\ldots,a_k\} \in \mathbb{B}$,

$$\begin{aligned}\varphi(B) &= \varphi(\{a_1,a_2,\ldots,a_k\})\\ &= (\varphi(a_1,a_2),\ldots,\varphi(a_1,a_k),\varphi(a_2,a_3),\ldots,\varphi(a_{k-1},a_k)).\\ &\qquad \varphi(a_i,a_j) \in Z_m, \quad 1 \le i < j \le k\end{aligned}$$

For an $RS_m(2,k,v)$, if there exists a mapping φ satisfying the following conditions:

(i) For each pair $\{a,b\} \subset Z_v$ with $a < b$, let $\{a,b\} \subset B_i$, $i = 1,2,\ldots,m$, let $\varphi(a,b)_i$ be the value of $\varphi(a,b)$ corresponding to the block B_i, then

$$\varphi(a,b)_i \equiv \varphi(a,b)_j \pmod m, \quad 1 \le i,j \le m$$

if and only if $i = j$.

(ii) For any block $B = \{a_1,a_2,\ldots,a_k\}$ with $0 \le a_1 < a_2 < \ldots < a_k \le v-1$

$$\varphi(a_r,a_s) + \varphi(a_s,a_t) \equiv \varphi(a_r,a_t) \pmod m \quad 1 \le r < s < t \le k.$$

Then we call it a labeled $RS_m(2,k,v)$ and denote the blocks in the following form:

$$(a_1,a_2,\ldots,a_k;\varphi(a_1,a_2),\ldots,\varphi(a_1,a_k),\varphi(a_2,a_3),\ldots,\varphi(a_{k-1},a_k)).$$

A labeled almost resolvable block design $ARS_m(2,k,v)$ can be defined in a similar way.

EXAMPLE 1. A labeled $RS_3(2,4,8)$:

{0,1,3,6;0,0,1,0,1,1}, {2,4,5,7;1,2,0,1,2,1};

{0,1,2,4;2,2,2,0,0,0}, {3,5,6,7;1,2,0,1,2,1};

{1,2,3,5;2,2,2,0,0,0}, {0,4,6,7;1,2,1,1,0,2};

{2,3,4,6;2,2,2,0,0,0}, {0,1,5,7;1,2,2,1,1,0};
{0,3,4,5;1,0,0,2,2,0}, {1,2,6,7;1,2,2,1,1,0};
{1,4,5,6;1,0,0,2,2,0}, {0,2,3,7;1,2,0,1,2,1};
{0,2,5,6;0,1,0,1,0,2}, {1,3,4,7;1,2,0,1,2,1}.

EXAMPLE 2. A labeled $ARS_3(2,4,5)$:

{0,1,2,3;1,1,0,0,2,2}, {1,2,3,4;1,1,0,0,2,2},
{0,2,3,4;0,1,1,1,1,0}, {0,1,3,4;2,2,0,0,1,1},
{0,1,2,4;0,2,2,2,2,0}.

2. LABELED ALMOST RESOLVABLE DESIGNS

Let K be a set of positive integers. A pairwise balanced design $S(2,K,v)$ is an ordered set $(X,\mathbb{B})$ where X is a finite set containing v elements, and $\mathbb{B}$ is a collection of subsets (called blocks) of X such that $|B| \in K$ for each block $B \in \mathbb{B}$ and each pair of distinct elements of X is contained in a unique block.

For a given set K, let

$$B(K) = \{v \mid \text{there exists an } S(2,K,v)\}.$$

K is called a PBD-closed set if $B(K) \subset K$.

If there exists an almost resolvable design $ARS_\lambda(2,k,v)$, then obviously we have

$$v \equiv 1 \pmod{k}$$

and the number of almost parallel classes is

$$\frac{\lambda\binom{v}{2}}{\binom{k}{2}\frac{v-1}{k}} = \frac{\lambda v}{k-1} .$$

For any $x \in X$, the number of almost parallel classes missing x is

$$\frac{\lambda v}{k-1} - \frac{\lambda(v-1)}{k-1} = \frac{\lambda}{k-1} .$$

Thus we have proved the following lemma.

LEMMA 1. If there exists an $ARS_\lambda(2,k,v)$, then

$$v \equiv 1 \pmod{k} \qquad (1)$$

$$\lambda \equiv 0 \pmod{(k-1)}. \qquad (2)$$

For given k and λ satisfying (2), let

$$LAB^*(k) = \{v \mid \text{there exists a labeled } ARS_\lambda(2,k,v)\}.$$

THEOREM 1. If $\lambda \equiv 0 \pmod{(k-1)}$, then $LAB^*(k)$ is a PBD-closed set.

PROOF. Suppose $\lambda = t(k-1)$. Let $(X,\mathbb{B})$ be an $S(2,K,v)$ where

$$K = \{k_1,k_2,\ldots,k_s\} \subseteq LAB^*(k).$$

For each $B \in \mathbb{B}$, $|B| = k_i$, form a labeled $ARS_\lambda(2,k,k_i)$ on B. For each $x \in B$, denote the t almost parallel classes missing x by $B_x^1, B_x^2, \ldots, B_x^t$. Then we construct a labeled $ARS_\lambda(2,k,v)$ on X as follows: For each $x \in X$, the t almost parallel classes missing x are:

$$\bigcup_{x\in B} B_x^1,\ \bigcup_{x\in B} B_x^2,\ \ldots,\ \bigcup_{x\in B} B_x^t .$$

This completes the proof.

Now let us consider the simplest case, i.e. the existence of labeled $ARS_2(2,3,v)$. If there exists a labeled $ARS_2(2,3,v)$, then

$$v \equiv 1 \pmod 3$$

and in each block the number of labelings 1 must be 0 or 2. So the following is a necessary condition for the existence of a labeled $ARS_2(2,3,v)$:

$$v \equiv 1 \text{ or } 4 \pmod{12} \tag{3}$$

We will prove that this condition is also sufficient.

THEOREM 2. The necessary and sufficient condition for the existence of a labeled $ARS_2(2,3,v)$ is

$$v \equiv 1 \text{ or } 4 \pmod{12}.$$

PROOF. We construct a labeled $ARS_2(2,3,4)$ as follows:

(0,1,2;0,1,1), (0,1,3;1,1,0), (0,2,3;0,0,0), (1,2,3;0,1,1).

It is well known that ([3])

$$B(4) = \{v \mid v \equiv 1,\ 4 \pmod{12}\}. \tag{4}$$

By Theorem 1, $LAB_2^*(3)$ is a PBD-closed set. Thus for every $v \in B(4)$ there exists a labeled $ARS_2(2,3,v)$. The conclusion then follows.

THEOREM 3. If $v \equiv 1 \pmod 4$ is a prime power, then there exists a labeled $ARS_3(2,4,v)$.

PROOF. Let $v = 4t+1$ and let α be a fixed primitive element of the Galois field $GF(v)$. Let $X = GF(v)$,

$$\mathbb{B} = \{B_{ij} \mid i = 0,1,\ldots,t-1;\ j = 0,1,\ldots,v-1\}$$

where

$$B_{i0} = \{\alpha^i, \alpha^{i+t}, \alpha^{i+3t}, \alpha^{i+2t}\}, \quad i = 0,1,\ldots,t-1$$
$$B_{ij} = \{\alpha^i+\alpha^j,\ \alpha^{i+t}+\alpha^j,\ \alpha^{i+3t}+\alpha^j,\ \alpha^{i+2t}+\alpha^j\},$$
$$i = 0,1,\ldots,t-1;\ j = 1,2,\ldots,v-1.$$

Then $(X,\mathbb{B})$ is an $S_3(2,4,v)$ ([3]). It is obviously almost resolvable.

As $v \equiv 1 \pmod 4$ and $t \equiv (v-1)/4$, we have

$$\alpha^{2t} - \alpha^{3t} = \alpha^t - 1$$
$$\alpha^{2t} - \alpha^t = \alpha^{3t} - 1 = \alpha^t(\alpha^t-1)$$
$$\alpha^{2t} - 1 = -2$$
$$\alpha^{3t} - \alpha^t = -2\alpha^t$$

So we can form a labeled $ARS_3(2,4,v)$ on $GF(v)$ with blocks as follows:

$$B_{i0} = (\alpha^i, \alpha^{i+t}, \alpha^{i+3t}, \alpha^{i+2t}; 1,1,0,0,2,2) \quad i = 0,1,\ldots,t-1$$
$$B_{ij} = (\alpha^i+\alpha^j,\ \alpha^{i+t}+\alpha^j,\ \alpha^{i+3t}+\alpha^j,\ \alpha^{i+2t}+\alpha^j; 1,1,0,0,2,2)$$
$$i = 0,1,\ldots,t-1,\ j = 1,2,\ldots,v-1.$$

3. LABELED RESOLVABLE BLOCK DESIGNS

Let $(X_1,\mathbb{B}_1)$ be an $S_\lambda(2,k,v_1)$ and $(X,\mathbb{B})$ be an $S_\lambda(2,k,v)$. If $X_1 \subseteq X$ and $\mathbb{B}_1 \subseteq \mathbb{B}$, then $(X_1,\mathbb{B}_1)$ is called a sub-$S_\lambda(2,k,v_1)$ of $(X,\mathbb{B})$.

Let $(X_1,\mathbb{B}_1)$ be a sub-$S_\lambda(2,k,v_1)$ of $(X,\mathbb{B})$. Further, suppose that $(X_1,\mathbb{B}_1)$ is an $RS_\lambda(2,k,v_1)$, $(X,\mathbb{B})$ is an $RS_\lambda(2,k,v)$ and each parallel class of $(X_1,\mathbb{B}_1)$ is a part of some parallel class of $\mathbb{B}$. Then $(X_1,\mathbb{B}_1)$ is called a sub-$RS_\lambda(2,k,v_1)$ of $(X,\mathbb{B})$.

Let $(X_1,\mathbb{B}_1)$ be a labeled $RS_\lambda(2,k,v_1)$ with mapping

$$\varphi_1: \mathbb{B}_1 \longrightarrow \{Z_\lambda\}^{\binom{k}{2}}$$

and $(X,\mathbb{B})$ be a labeled $RS_\lambda(2,k,v)$ with mapping

$$\varphi: \mathbb{B} \longrightarrow \{Z_\lambda\}^{\binom{k}{2}}.$$

If $(X_1,\mathbb{B}_1)$ is a sub-$RS_\lambda(2,k,v_1)$ of $(X,\mathbb{B})$ and φ_1 is the restriction of φ to $\mathbb{B}_1$, then $(X_1,\mathbb{B}_1)$ is called a sub-labeled $RS_\lambda(2,k,v_1)$ of $(X,\mathbb{B})$.

In this section we will give constructions for labeled resolvable block designs.

Let $K = \{k_1,\ldots,k_s\}$ and $M = \{m_1,\ldots,m_t\}$ be two sets of positive integers. A group divisible design GD(K,M;v) is an ordered triple $(X,\mathbb{G},\mathbb{B})$ where X is a finite set containing v elements, $\mathbb{G}$ is a collection of subsets (called groups) of X which partitions X and for each $G \in \mathbb{G}$, $|G| \in M$; $\mathbb{B}$ is a collection of subsets (called blocks) of X such that for each $B \in \mathbb{B}$, $|B| \in K$ and each pair of elements of X from distinct groups occurs in a unique block.

THEOREM 4. Let $K = \{k_1,\ldots,k_s\}$, $M = \{m_1,\ldots,m_r\}$. Let $(X_0,\mathbb{B}_0)$ be a labeled $RS_{k-1}(2,k,v_0)$. If there exists a GD(K,M;v) such that:

(i) for each $m_i \in M$, there exists a labeled $RS_{k-1}(2,k,m_i+v_0)$ containing a sub-labeled $RS_{k-1}(2,k,v_0)$;

(ii) for each $k_j \in K$, there exists a labeled $ARS_{k-1}(2,k,k_j)$,

then there exists a labeled $RS_{k-1}(2,k,v+v_0)$ which contains a sub-labeled $RS_{k-1}(2,k,v_0)$ and contains a sub-labeled $RS_{k-1}(2,k,m_i+v_0)$ for each $m_i \in M$.

PROOF. First we note that in an $RS_{k-1}(2,k,v)$, the number of parallel classes is v-1, and in an $ARS_{k-1}(2,k,v)$, the number of almost parallel classes is v.

Denote the v_0-1 parallel classes of the labeled $RS_{k-1}(2,k,v_0)$ by $\pi_1,\pi_2,\ldots,\pi_{v_0-1}$. For each $G \in \mathbb{G}$, $|G| = m_i$, form a labeled $RS_{k-1}(2,k,m_i+v_0)$ containing $(X_0,\mathbb{B}_0)$ as a sub-labeled $RS_{k-1}(2,k,v_0)$, and denote the parallel class containing π_t by G_t for $t = 1,\ldots,v_0-1$ and write $G_t' = G_t\backslash\pi_t$. The remaining m_i parallel classes contain no blocks of $\mathbb{B}_0$; we denote those parallel classes by a_x^G in an arbitrary way for each $x \in G$. For each block $B \in \mathbb{B}$, $|B| = k_j$, form a labeled $ARS_{k-1}(2,k,k_j)$ on B, and for each $x \in B$, denote the almost parallel class missing x by a_x^B. In this way we obtain a labeled $RS_{k-1}(2,k,v+v_0)$ on $X \cup X_0$. The $v+v_0-1$ parallel classes are

$$\pi_t \cup (\bigcup_{G\in\mathbb{G}} G_t'), \quad t = 1,2,\ldots,v_0-1$$

and

$$a_x^G \cup \left\{\bigcup_{x\in B} a_x^B\right\}, \quad x \in X,$$

where for each $x \in X$, G is the unique group containing x.

If we let $v_0 = 1$ in Theorem 4, then we obtain the following corollary:

COROLLARY. Let $K = \{k_1,\ldots,k_s\}$, $M = \{m_1,\ldots,m_r\}$. If there exists a GD(K,M;v) such that:

(i) for each $m_i \in M$, there exists a labeled $RS_{k-1}(2,k,m_i+1)$,

(ii) for each $k_j \in K$, there exists a labeled $ARS_{k-1}(2,k,k_j)$,

then there exists a labeled $RS_{k-1}(2,k,v+1)$ which contains a sub-labeled $RS_{k-1}(2,k,m_i+1)$ for each $m_i \in M$.

EXAMPLE 3. The following are a labeled $ARS_2(2,3,4)$ and a labeled $RS_2(2,3,9)$:

$ARS_2(2,3,4)$:
(0,1,2;0,0,0), (0,1,3;1,0,1), (0,2,3;1,1,0), (1,2,3;1,0,1).

$RS_2(2,3,9)$:

(0,1,2;0,0,0), (3,4,5;0,0,0), (6,7,8;0,0,0)
(0,3,6;1,1,0), (1,4,7;0,1,1), (2,5,8;1,0,1)

(0,4,8;0,0,0), (1,5,6;1,0,1), (2,3,7;1,0,1)
(0,5,7;1,0,1), (1,3,8;1,1,0), (2,4,6;0,1,1)
(0,4,7;1,1,0), (1,2,8;1,0,1), (3,5,6;1,1,0)
(0,3,5;1,0,1), (1,3,7;0,0,0), (4,6,8;0,1,1)
(0,3,8;0,1,1), (1,4,5;1,0,1), (2,6,7;0,1,1)
(0,1,6;1,0,1), (2,3,4;0,1,1), (5,7,8;0,1,1)

Thus we can construct by Theorem 4 a labeled $RS_2(2,3,33)$ from a GD(4,8;32) (i.e. a transversal design TD[4,8]).

It is not difficult to prove the following generalization of Theorem 4.

THEOREM 5. Suppose $\lambda \equiv 0 \pmod{(k-1)}$. Let K and M be two sets of positive integers. Let $(X_0,\mathbb{B}_0)$ be a labeled $RS_\lambda(2,k,v_0)$. If there exists a GD(K,M;v) such that:

(i) for each $m_i \in M$, there exists a labeled $RS_\lambda(2,k,m_i+v_0)$ containing $(X_0,\mathbb{B}_0)$ as a sub-labeled $RS_\lambda(2,k,v_0)$,

(ii) for each $k_j \in K$, there exists a labeled $ARS_\lambda(2,k,k_j)$, then there exists a labeled $RS_\lambda(2,k,v+v_0)$ containing a sub-labeled $RS_\lambda(2,k,v_0)$ and a sub-labeled $RS_\lambda(2,k,m_i+v_0)$ for each $m_i \in M$.

4. RESOLVABLE GROUP DIVISIBLE DESIGNS

If K = {k} and M = {m}, then a group divisible design GD({k},{m};v) is simply denoted GD(k,m;v). A GD(k,m,v) is called resolvable and denoted RGD(k,m;v) if the blocks can be partitioned into parallel classes.

From the above definition, a Kirkman system RS(2,k,v) is an RGD(k,1;v) if we take each point as a group, or an RGD(k,k;v) if we take the blocks of a fixed parallel class as groups.

An RGD(k,k-1;v) is also called a nearly Kirkman system. In particular, an RGD(3,2;v) is called a nearly Kirkman triple system.

For fixed k and m, to determine the spectrum of v for which there is an RGD(k,m;v) is a fundamental and difficult problem. It is not difficult to show that the following are necessary conditions for the existence of an RGD(k,m;v):

$$v \equiv 0 \pmod{k}, \quad v \equiv 0 \pmod{m}$$
$$v-m \equiv 0 \pmod{(k-1)}. \tag{9}$$

The problem is: are the necessary conditions (9) also sufficient for the existence of an RGD(k,m;v)?

For k = m = 3, this is the famous Kirkman's schoolgirl problem. It was completely solved by D. K. Ray-Chaudhuri and R. M. Wilson [7]: there exists a Kirkman triple system RGD(3,2;v) if and only if

$$v \equiv 3 \pmod{6}. \tag{10}$$

For k = m = 4, H. Hanani et al [4] proved that there exists a Kirkman system RS(2,4,v) if and only if

$$v \equiv 4 \pmod{12}. \tag{11}$$

For k = 3, m = 2, it was proved ([1],[2],[6], see also [5]) that there exists a nearly Kirkman triple system RGD(3,2;v) if and only if

$$v \equiv 0 \pmod{6} \quad v > 12. \tag{12}$$

R. Rees and D. R. Stinson [8] considered the general case of RGD(3,m;v) and proved that the necessary conditions (9) are also sufficient with three exceptions (m = 2, v = 6,12 and m = 6, v = 18) and a few possible exceptions.

The present author obtained some direct and recursive constructions for RGD(k,m,v) in [10] and proved in [11] that the necessary and sufficient condition for the existence of a nearly Kirkman system RGD(4,3;v) is

$$v \equiv 0 \pmod{12} \tag{13}$$

with the exception v = 12 and fifteen other possible exceptions.

In the remainder of this section we will discuss applications of labeled resolvable block designs to the construction of resolvable group divisible designs.

THEOREM 6. If there exists a labeled $RS_m(2,k;v)$, then there exists an RGD(k,m;mv).

PROOF. Let $(I_v,\mathbb{B})$ be a labeled $RS_m(2,k,v)$. Let each $a_i \in I_v$ correspond to an m-set $\{a_{i0},a_{i1},\dots,a_{i,m-1}\}$. To construct an

RGD(k,m;mv), we take $\{a_0,a_1,\ldots,a_{m-1}\}$ as a group, for each $a \in I_v$. Let

$$B = (a_0,a_1,\ldots,a_{k-1};\ \varphi(a_0,a_1),\ldots,\varphi(a_0,a_{k-1}),\varphi(a_1,a_2),\ldots,\varphi(a_{k-2},a_{k-1}))\quad \text{where}\quad a_0 < a_1 < \ldots < a_{k-1}.$$

If

$$\varphi(a_0,a_i) = s_i \qquad i = 1,2,\ldots,k-1.$$

Then as φ is a mapping satisfying the conditions (i) and (ii), we have

$$\varphi(a_i,a_j) \equiv \varphi(a_0,a_j)-\varphi(a_0,a_i) \equiv s_j-s_i \pmod{m}$$

for $0 < i < j \leq k-1$. So we may take

$$\{a_{0,j},a_{1,j+s_1},\ldots,a_{k-1,j+s_{k-1}}\}, \qquad j \equiv 0,1,\ldots,m-1 \pmod{m}$$

as blocks. Thus from each parallel class of the labeled $RS_m(2,k,v)$ we obtain a parallel class and it can be verified that these are the $m(v-1)/(k-1)$ parallel classes of an RGD(k,m;mv) with

$$\{a_{i0},a_{i1},\ldots,a_{i,m-1}\}, \qquad a_i \in I_v$$

as groups.

We hope to use the above theorem to construct RGD(k,m;v) for small v.

EXAMPLE 4. We have constructed a labeled $RS_2(2,3,9)$ in Example 3; thus we obtain a nearly Kirkman triple system RGD(3,2;18). We note that 18 is the least v for which there exists an RGD(3,2;v). We also have obtained a labeled $RS_3(2,4,8)$ in Example 1; thus we obtain a nearly Kirkman system RGD(4,3;24) and 24 is the least v for which there exists an RGD(4,3;v).

The following Corollary is useful in the construction of nearly Kirkman systems.

COROLLARY. If there exists a labeled $RS_{k-1}(2,k;v)$, then there exists a nearly Kirkman system RGD(k,k-1;(k-1)v).

We may also define labeled BIB designs. Constructions of labeled BIB designs and the applications in the construction of group divisible designs and BIB designs will be discussed in a subsequent paper.

ACKNOWLEDGEMENT

I got the idea for labeled resolvable triple systems while I was visiting the University of Toronto. I would like to thank Professor Eric Mendelsohn and Professor Zhu Lie for their valuable discussions.

REFERENCES

[1] R. D. Baker and R. M. Wilson, Nearly Kirkman Triple Systems. Utilitas Math. 11(1977), 289-296.

[2] A. E. Brouwer, Two new nearly Kirkman triple systems. Utilitas Math. 13(1978), 311-314.

[3] H. Hanani, On balanced incomplete block designs and related designs. Discrete Math. 11(1975), 255-369.

[4] H. Hanani, D. K. Ray-Chaudhuri and R. M. Wilson, On resolvable designs. Discrete Math. 3(1972), 343-357.

[5] C. Huang, E. Mendelsohn and A. Rosa, On partially resolvable t-partitions. Ann. Discrete Math. 12(1982), 169-183.

[6] A. Kotzig and A. Rosa, Nearly Kirkman Systems, Proc. 5th Southeastern Conference on Combinatorics, Graph Theory, and Computing (Boca Raton, 1974), 607-614.

[7] D. K. Ray-Chaudhuri and R. M. Wilson, Solution of Kirkman's schoolgirl problem, Proc. Symp. Pure Math. 19 (American Math. Soc., 1971), 187-203.

[8] R. Rees and D. R. Stinson, On the existence of Kirkman triple systems containing Kirkman subsystems. Ars. Combinatoria 26(1988), 3-16.

[9] Shen Hao, Constructions of resolvable group divisible designs with block size 4. J. Comb. Math. Comb. Comput. 1(1987), 145-150.

[10] Shen Hao, On resolvable group divisible designs. Preprint Series in Mathematics (University of Toronto, 1987), No. 1.

[11] Shen Hao, On nearly Kirkman systems with block size 4. Preprint.

Existence of Simple Resolvable Block Designs with Block Size 4

SHEN Hao, Department of Applied Mathematics, Shanghai Jiao Tong University, Shanghai 200030, People's Republic of China

WU Minzhen, Department of Applied Mathematics, Shanghai Jiao Tong University, Shanghai 200030, People's Republic of China

1. INTRODUCTION

A balanced incomplete block design $S_\lambda(2,k,v)$ is an ordered pair $(V,\mathcal{B})$ where V is a finite set containing v elements and $\mathcal{B}$ is a collection of k-subsets (called blocks) of V such that each pair of distinct elements is contained in exactly λ blocks.

Let $(V,\mathcal{B})$ be an $S_\lambda(2,k,v)$. A parallel class is a set of blocks which partitions V. $(V,\mathcal{B})$ is called resolvable if $\mathcal{B}$ can be partitioned into parallel classes. A resolvable $S_\lambda(2,k,v)$ will be denoted $RS_\lambda(2,k,v)$. An almost parallel class is a set of blocks which partitions $V\setminus\{x\}$ for some $x \in V$. $(V,\mathcal{B})$ is called almost resolvable if $\mathcal{B}$ can be partitioned into almost parallel classes. An almost resolvable $S_\lambda(2,k,v)$ will be denoted $ARS_\lambda(2,k,v)$.

It can be easily seen that the following condition

$$v \equiv k\left[\bmod \frac{k(k-1)}{\lambda_0}\right], \quad \lambda_0 = (\lambda,k-1). \tag{1}$$

is necessary for the existence of an $RS_\lambda(2,k,v)$. It is well known ([11]) that for $\lambda = 1$ and $k = 3$ or 4, there is an $RS_1(2,k,v)$ if and only if $v \equiv k \pmod{k(k-1)}$.

It has been proved ([10]) that there is an $RS_2(2,3,v)$ if and only if $v \equiv 0 \pmod 3$, $v \neq 6$, and ([1]) there is an $RS_3(2,4,v)$ if and only if $v \equiv 0 \pmod 4$. But their constructions admit repeated blocks.

In some cases, we are interested in designs without repeated blocks. Such designs will be called simple. For $\lambda \geq 2$ the existence of simple $S_\lambda(2,k,v)$ has been studied by several authors ([4],[5],[6],[8], [14]).

But there are not many known results concerning the existence of simple $RS_\lambda(2,k,v)$. Recently, it was proved ([13]) that there exists a simple $RS_2(2,3,v)$ if and only if

$$v \equiv 0 \pmod 3, \quad v > 6. \tag{2}$$

In this paper, we will consider the existence and construction of simple $RS_\lambda(2,4,v)$ and prove that for $\lambda = 2$, 4 and 5, the obvious necessary conditions for the existence of a simple $RS_\lambda(2,4,v)$ are also sufficient, and that the necessary conditions for the existence of a simple $RS_3(2,4,v)$ are also sufficient with 2 possible exceptions.

2. GENERAL CONSTRUCTIONS

A transversal design $TD[k,\lambda;n]$ is an ordered triple $(V,\mathcal{G},\mathcal{B})$ where V is a v-set, $v = kn$, $\mathcal{G}$ is a collection of n-subsets (called groups) of V which partitions V, and $\mathcal{B}$ is a collection of k-subsets (called blocks) such that each block intersects each group in a unique point and each pair of points from distinct groups is contained in exactly λ blocks. A $TD[k,\lambda;n]$ is called resolvable and denoted $RTD[k,\lambda;n]$ if the blocks can be partitioned into parallel classes. A $TD[k,\lambda;n]$ is called simple if it contains no repeated blocks. It is well known that the existence of an $RTD[k,1;n]$ is equivalent to the existence of $k-1$ pairwise orthogonal Latin squares of order n.

LEMMA 1. If $n \geq 4$, $n \neq 6,10$, then there exists a simple

RTD[4,3;n].

PROOF. If $n \geq 4$, $n \neq 6,10$, then there exist 3 pairwise orthogonal Latin squares of order n ([3]) and then there exists an RTD[4,1;n]. Let A,B,C and D be 4 pairwise disjoint n-sets, and $A = Z_n$, $V = A \cup B \cup C \cup D$, and $(V,\mathcal{G},\mathcal{B})$ be an RTD[4,1;n] where $\mathcal{G} = \{A,B,C,D\}$. Let $\mathcal{B}_i = \{\{a+i,b,c,d\} \mid \{a,b,c,d\} \in \mathcal{B}, (a,b,c,d) \in A\times B\times C\times D\}$, $i = 1,2$. Then $(V,\mathcal{G},\mathcal{B} \cup \mathcal{B}_1 \cup \mathcal{B}_2)$ is an RTD[4,3;n] without repeated blocks.

LEMMA 2. If there exists a simple $ARS_3(2,4,v)$ or a simple $RS_3(2,4,v)$, then there exists a simple $RS_3(2,4,4v)$.

PROOF. Suppose there is a simple $RS_3(2,4,v)$ (or a simple $ARS_3(2,4,v)$). Then $v \equiv 0 \pmod 4$ (or $v \equiv 1 \pmod 4$). So by Lemma 1 there exists a simple RTD[4,3;v] with $\{i\}\times I_v$ as groups, $i \in Z_4$ and $I_v = \{1,2,\ldots,v\}$. Without loss of generality, let the following be one of the parallel classes:

$$B_0: \{(0,1),(1,1),(2,1),(3,1)\}, \{(0,2),(1,2),(2,2),(3,2)\},\ldots\{(0,v),(1,v),(2,v),(3,v)\}$$

The remaining parallel classes are denoted by $\mathcal{B}_1,\mathcal{B}_2,\ldots,\mathcal{B}_{3v-1}$.

i) Suppose there is a simple $RS_3(2,4,v)$. For each $i \in Z_4$, let $\mathcal{B}_{ij}$, $j = 1,2,\ldots,v-1$, be the parallel classes of a simple $RS_3(2,4,v)$ on $\{i\}\times I_v$. Then the parallel classes of the desired simple $RS_3(2,4,4v)$ on $Z_4\times I_v$ are $\mathcal{B}_s$, $s = 0,1,\ldots,3v-1$ and $\bigcup_{i\in Z_4} \mathcal{B}_{ij}$, $j = 1,2,\ldots,v-1$,

ii) Suppose there is a simple $ARS_3(2,4,v)$. Then for each $i \in Z_4$ form a simple $ARS_3(2,4,v)$ on $\{i\}\times I_v$, and denote the almost parallel class missing (i,j) by $\mathcal{B}_{ij}$, $j \in I_v$. Then the parallel classes of the desired simple $RS_3(2,4,4v)$ on $Z_4\times I_v$ are $\mathcal{B}_s$, $s = 1,2,\ldots,3v-1$, and

$$(\bigcup_{i\in Z_4} \mathcal{B}_{ij}) \cup \{(0,j),(1,j),(2,j),(3,j)\}, \quad j \in I_v.$$

It is proved ([2]) that if there is an $RS_3(2,4,u)$ or there is an $ARS_3(2,4,v)$, then there exists an $RS_3(2,4,uv)$. With the existence of a simple RTD[4,3;v] for every $v \equiv 1 \pmod 4$, $v \geq 5$, we can prove the following stronger result.

LEMMA 3. If there exist a simple $RS_3(2,4,u)$ and a simple $RS_3(2,4,v)$ or $ARS_3(2,4,v)$, then there exists a simple $RS_3(2,4,uv)$. If there exist a simple $RS_3(2,4,m+1)$ and a simple $ARS_3(2,4,u)$, then there exists a simple $RS_3(2,4,mu+1)$.

LEMMA 4. If $v \equiv 1 \pmod 4$ is a prime power, then there exists a simple $ARS_3(2,4,v)$.

PROOF. Let $v = 4t+1$ be a prime power and α be a primitive element of $GF(v)$. The following construction of $ARS_3(2,4,v)$ can be found in [2], here we list only the base blocks,

$$\{ \alpha^0, \alpha^t, \alpha^{2t}, \alpha^{3t} \}$$
$$\{ \alpha^1, \alpha^{t+1}, \alpha^{2t+1}, \alpha^{3t+1} \}$$
$$\{ \alpha^2, \alpha^{t+2}, \alpha^{2t+2}, \alpha^{3t+2} \}$$
$$\{ \alpha^3, \alpha^{t+3}, \alpha^{2t+3}, \alpha^{3t+3} \}$$

It can be verified that this $ARS_3(2,4,v)$ is both simple and almost resolvable.

Let K and M be two sets of positive integers. A group divisible design $GD(K,M;v)$ is a triple $(V,\mathcal{G},\mathcal{A})$ where V is a v-set, $\mathcal{G}$ is a set of subsets (called groups) of V which partitions V and $|G| \in M$ for each $G \in \mathcal{G}$, and $\mathcal{A}$ is a collection of subsets (called blocks) of V such that $|B| \in K$ for each $B \in \mathcal{A}$ and each pair of elements of V from distinct groups is contained in a unique block.

A construction of simple $RS_2(2,3,v)$ by using group divisible designs was given in [13]. Here we will prove a similar construction for simple $RS_3(2,4,v)$.

LEMMA 5. If there is a $GD(K,M;v)$ such that

i) for each $m \in M$, there exists a simple $RS_3(2,4,m+1)$.

ii) for each $k \in K$, there exists a simple $ARS_3(2,4,k)$.

Then there exists a simple $RS_3(2,4,v+1)$.

PROOF. For each group $G \in \mathcal{G}$, $|G| = m$, form a simple $RS_3(2,4,m+1)$ on $G \cup \{\infty\}$ where ∞ is a new element, and denote the parallel classes by a_x^G for each $x \in G$. For each block $B \in \mathcal{A}$, $|B| = k$, form a simple $ARS_3(2,4,k)$ and denote the almost parallel class missing x by a_x^B for each $x \in B$. Then we obtain a simple $RS_3(2,4,v+1)$ on $V \cup \{\infty\}$, and for each $x \in V$, the parallel class a_x is the union of a_x^G, where G is the unique group containing x, with all the almost parallel classes a_x^B, where $B \in \mathcal{A}$ is any block containing x.

For construction of simple $RS_\lambda(2,4,v)$, the following result is also needed.

LEMMA 6. ([7]) Let T_1 and T_2 be two sets of k-subsets of V. $|V| = v$, and $|T_1| = b_1$, $|T_2| = b_2$. If $b_1 b_2 < \binom{v}{k}$, then there exists a permutation π on V such that $\pi(T_1) \cap T_2 = \emptyset$.

3. EXISTENCE OF SIMPLE $RS_\lambda(2,4,v)$ FOR $\lambda = 2,4$ AND 5

For $\lambda = 2,4$ and 5, the necessary conditions for the existence of a simple $RS_\lambda(2,4,v)$ is $v \equiv 4 \pmod{12}$, $v > 4$.

The purpose of this section is to prove that this condition is also sufficient.

LEMMA 7. If there exist a simple $RS_{\lambda_1}(2,4,v)$ and a simple $RS_{\lambda_2}(2,4,v)$ such that $\lambda_1\lambda_2\left[\frac{v(v-1)}{12}\right]^2 < \binom{v}{4}$, then there exists a simple $RS_{\lambda_1+\lambda_2}(2,4,v)$.

PROOF. Let $(V,\mathcal{B}_1)$ be a simple $RS_{\lambda_1}(2,4,v)$ and $(V,\mathcal{B}_2)$ be a simple $RS_{\lambda_2}(2,4,v)$. In Lemma 6, let $T_1 = \mathcal{B}_1$, $T_2 = \mathcal{B}_2$. Then $b_1 = \frac{\lambda_1 v(v-1)}{12}$, $b_2 = \frac{\lambda_2 v(v-1)}{12}$. Then it can be easily seen that

$$b_1 b_2 = \lambda_1\lambda_2\left[\frac{v(v-1)}{12}\right]^2 < \binom{v}{4}.$$

Thus by Lemma 6, there exists a permutation π such that $\pi(\mathcal{B}_1) \cap \mathcal{B}_2 = \emptyset$. So $(V,\pi(\mathcal{B}_1) \cup \mathcal{B}_2)$ is a simple $RS_{\lambda_1+\lambda_2}(2,4,v)$.

THEOREM 1. There exists a simple $RS_\lambda(2,4,v)$ for $\lambda = 2,4$ or 5 if and only if $v \equiv 4 \pmod{12}$, $v \geq 16$.

PROOF. It is well known that there exists an $RS_1(2,4,v)$ for every $v \equiv 4 \pmod{12}$. Obviously if $\lambda_1\lambda_2 \leq 4$ and $v \geq 16$ then $\lambda_1\lambda_2\left[\frac{v(v-1)}{12}\right]^2 < \binom{v}{4}$. So letting $\lambda_1 = \lambda_2 = 1$, we obtain a simple $RS_2(2,4,v)$ from an $RS_1(2,4,v)$ for every $v \equiv 4 \pmod{12}$, $v \geq 16$. Then letting $\lambda_1 = \lambda_2 = 2$, we obtain a simple $RS_4(2,4,v)$ from a simple $RS_2(2,4,v)$ for every $v \equiv 4 \pmod{12}$, $v \geq 16$. To complete the proof of the theorem, let $\lambda_1 = 1$, $\lambda_2 = 4$. We obtain a simple $RS_5(2,4,v)$ from an $RS_1(2,4,v)$ and a simple $RS_4(2,4,v)$ for every $v \equiv 4 \pmod{12}$, $v \geq 16$.

4. EXISTENCE OF SIMPLE $RS_3(2,4,v)$

In this section we shall prove that the necessary and sufficient condition for the existence of a simple $RS_3(2,4,v)$ is $v \equiv 0 \pmod 4$, $v \geq 8$, with 2 possible exceptions.

For a given set K of positive integers, let

$$B(K) = \{v \mid \text{a } (v,K,1)\text{-PBD exists}\}$$

K is called PBD-closed if $B(K) = K$.

Let $AB_3^*(4) = \{v \mid \text{a simple } ARS_3(2,4,v) \text{ exists}\}$. We have the following result:

LEMMA 8. $AB_3^*(4)$ is PBD-closed.

PROOF. Let $(V,\mathcal{B})$ be a $(v,K,1)$-PBD, $K \subset AB_3^*(4)$. For each $B \in \mathcal{B}$, $|B| = k$, form a simple $ARS_3(2,4,k)$. For each $x \in B$, let B_x denote the almost parallel class missing x. Then we obtain a simple $ARS_3(2,4,v)$ on V. For each $x \in V$, the almost parallel class missing x is the union of all B_x, where $B \in \mathcal{B}$ is any block containing x.

The following lemma can be found in [12].

LEMMA 9. Let $K = \{5,9,13,17,29,49\}$. Then there exists a $(v,K,1)$-PBD for every positive integer $v \equiv 1 \pmod 4$ except possibly 33,57,93 and 133.

LEMMA 10. There exists a simple $RS_3(2,4,4v)$ for every $v \equiv 1 \pmod 4$ except possibly v = 33,57,93 and 133.

PROOF. By Lemma 4, there exists a simple $ARS_3(2,4,v)$ for v = 5,9,13,17,29,49. So with the above Lemma 9 and the PBD-closedness of $AB_3^*(4)$, there exists a simple $ARS_3(2,4,v)$ for every $v \in 1 \pmod 4$, $v \neq 33,57,93,133$. Then by Lemma 2, there exists a simple $RS_3(2,4,4v)$ for every $v \equiv 1 \pmod 4$, $v \neq 33,57,93,133$.

By Lemma 7 from the existence of a simple $RS_\lambda(2,4,v)$ for all $v \equiv 4 \pmod{12}$, $v \geq 8$, and $\lambda = 1,2$, we can prove the following lemma.

LEMMA 11. There exists a simple $RS_3(2,4,v)$ if $v \equiv 4 \pmod{12}$.

LEMMA 12. There exists a simple $RS_3(2,4,v)$ for every $v \in H_1$, where $H_1 = \{8,12,24,132,152,228,264\}$.

PROOF. For each $v \in H_1$, an $RS_3(2,4,v)$ can be found in ([1]); it can be checked that for any $v \in H_1$, the $RS_3(2,4,v)$ constructed there is simple.

LEMMA 13. There exists a simple $RS_3(2,4,v)$ for every $v \in H_2$, where $H_2 = \{48,80,96,112,128,144,192,272,320\}$.

PROOF. Each $v \in H_2$ can be written in the form v = 4m, (see Table 1). By Lemma 2, there exists a simple $RS_3(2,4,v)$ for each $v \in H_2$.

Table 1

v=4m	m	v=4m	m
48	12	144	36
80	20	192	48
96	24	272	68
112	28	320	80
128	32		

LEMMA 14. There exists a simple $RS_3(2,4,v)$ for every $v \in H_3$, where $H_3 = \{56,176,236,248,276,796\}$.

PROOF. $56 = 5\cdot 11+1$, $176 = 5\cdot 35+1$, $236 = 5\cdot 47+1$, $248 = 13\cdot 19+1$, $276 = 5\cdot 55+1$, $796 = 5\cdot 159+1$. As there exist a simple $ARS_3(2,4,5)$ and a simple $ARS_3(2,4,13)$ and there exists a simple $RS_3(2,4,m+1)$ for $m = 11,19,35,47,159$, the conclusion then follows from Lemma 5 and the existence of a $TD[13,1;19]$ and a $TD[5,1;m]$ for each $m = 11,19,35,47$, 55 and 159.

LEMMA 15. There exists a simple $RS_3(2,4,v)$ for every $v \in H_4$, where $H_4 = \{60,72,104,108,120,140,156,168,200,204,216,224,252,260,288,296,300,312,780,784,792\}$.

PROOF. Each $v \in H_4$ can be written in the form $v = mn$ (see Table 2). Using Lemma 3, we obtain a simple $RS_3(2,4,v)$ for every $v \in H_4$.

From Lemmas 10 through 15, we obtain the following lemma.

LEMMA 16. For every $v \equiv 0 \pmod 4$, $8 \leq v \leq 324$, $v \notin \{44,284\}$, there exists a simple $RS_2(2,4,v)$.

PROOF. Apply Lemma 3 with $m = 7$, $u = 13$, and $m = 11$, $u = 17$ to obtain simple $RS_3(2,4,v)$ for $v = 92,188$. All the values of v are dealt with except $v = 240$, for which we write $240 = 4\times 60$ and apply Lemma 2.

Table 2

v=mn	m	n	v=mn	m	n
60	12	5	224	28	8
72	8	9	252	12	21
104	8	13	260	20	13
108	12	9	288	24	12
120	24	5	296	8	37
140	28	5	300	60	5
156	12	13	312	24	13
168	8	21	780	60	13
200	40	5	784	16	49
204	12	17	792	88	9
216	24	9			

Now we are in a position to prove our main theorem.

THEOREM 2. There exists a simple $RS_3(2,4,v)$ if and only if $v \equiv 0 \pmod 4$, $v \geq 8$ with 2 possible exceptions.

PROOF. For each $v \equiv 0 \pmod 4$, $8 \leq v \leq 324$, $v \neq 44,284$, the existence of a simple $RS_3(2,4,v)$ was proved in Lemma 16. The existence of a simple $RS_3(2,4,788)$ was proved in Lemma 10. The existence of a simple $RS_3(2,4,796)$ was proved in Lemma 14. The existence of a simple $RS_3(2,4,v)$ for $v = 780,784,792$ was proved in Lemma 15. It was proved ([2]) that if there is a TD[17,1;n], then there exists a GD(K,M;v) where $K = \{5,17\}$, $M = \{n,n+4m_1, n+4m_2\}$, $v = 17n+4(m_1+m_2)$, $0 \leq m_1,m_2 \leq n$. Let

$$n = 19,27,31,47,67,83,107,139,199,243,323,443,619, 863,1207,1679,$$

and

$$3^s,\ 3^{s-3}\cdot 31,\ 3^{s-4}\cdot 113,\ 3^{s-3}\cdot 47,\ 3^{s-3}\cdot 59,\ 3^{s-3}\cdot 79,\ 3^{s-3}\cdot 103, 3^{s-3}\cdot 127, 3^{s-1}\cdot 19, 3^{s-1}\cdot 23, 3^{s+2},\quad s = 7,9,11,\ldots$$

Then for each such n, choose m_1 and m_2 such that $0 \leq m_1,m_2 \leq n$ and $n,n+4m_1,n+4m_2 \neq 43, 283, 779, 783, 787, 791$ or 795. By Lemma 5, we can prove that there is a simple $RS_3(2,4,v)$ for every $v \equiv 0 \pmod 4$, $v \geq 324$.

ACKNOWLEDGEMENT. The authors would like to thank the referee for many valuable suggestions.

REFERENCES

1. R.D. Baker, Resolvable BIBD and SOLS, Discrete Math. 44(1983), 13-19.

2. R.D. Baker, Whist tournaments, Congressus Num. (1975), 89-100.

3. T. Beth, D. Jungnickel and H. Lenz, Design Theory (Bibliographisches Institute, Zurich, 1986), pp. 643-644.

4. J. Van Buggenhaut, On the existence of 2-designs $S_2(2,3,v)$ without repeated blocks, Discrete Math. 8(1974), 105-109.

5. J. Van Buggenhaut, Existence and constructions of 2-designs $S_3(2,3,v)$ without repeated blocks, J. Geometry 4(1974), 1-10.

6. M. Dehon, On the existence of 2-designs $S_2(2,3,v)$ without repeated blocks, Discrete Math. 43(1983), 155-171.

7. B. Ganter, J. Pelikan and L. Teirlinck, Small sprawling systems of equicardinal sets, Ars Combinatoria 4(1977), 133-142.

8. Guo Hai-tao, On the existence of B(4,2;v) and B(4,3;v) without repeated blocks J. Shanghai Jiao Tong University (to appear).

9. H. Hanani, Balanced incomplete block designs and related designs, Discrete Math. 11(1975), 255-369.

10. H. Hanani, On resolvable balanced incomplete block designs, J. Combinatorial Theory A 17A, 275-289.

11. H. Hanani, D.K. Ray-Chaudhuri and R.M. Wilson, On resolvable designs, Discrete Math. 3(1972), 343-357.

12. R.C. Mullin, P.J. Schellenberg, S.A. Vanstone and W.D. Wallis, On the existence of frames, Discrete Math. 37(1981), 79-104.

13. Shen Hao, Resolvable twofold triple systems without repeated blocks Kexue Tongbao (Science Bulletin) 33(1988), 1855-1857.

14. D.R. Stinson, W.D. Wallis, Twofold triple systems without repeated blocks, Discrete Math. 47(1983), 125-128.

A Special Class of Block Designs

SUN Shixin, Department of Computer Science, Chengdu Institute of Radio Engineering, Chengdu, China

KANG Tai, Department of Mathematics, Chengdu Teachers College, Chengdu, China

1. INTRODUCTION

Let $S = \{X_1, X_2, \ldots, X_n\}$ be a finite set. We consider the following problem: does there exist a block design $B = \{B_1, B_2, \ldots, B_N\}$ on S such that

a) $n = C_N^2 = N(N-1)/2$

b) If f_{ij} $(i \neq j)$ denotes the number of places in which rows i and j of the incidence matrix of B differ, then the f_{ij} are distinct and $\{f_{ij} \mid i \neq j\} = \{1, 2, \ldots, n\}$.

The hypercube Q_n is defined to be the graph whose vertex set $V(Q_n)$ is $\{0,1\}^n$. For $a_i, a_j \in V(Q_n)$, $a_i a_j \in E(Q_n)$ iff a_i and a_j differ in exactly one coordinate.

For $a_i, a_j \in V(Q_n)$, the Hamming distance $d(a_i, a_j)$ is defined to be the number of coordinates in which a_i and a_j differ from each other. It should be noted that the Hamming distance coincides with the graph theoretic distance; i.e. the length of a shortest $(a_i - a_j)$ path.

Our original problem is closely related to the following: How can one find a maximal subset $\{a_1, a_2, \ldots, a_N\}$ of vertices in the hypercube Q_n such that all the (Hamming) distances $d(a_i, a_j)$ are different. In other words,

$$d(a_i, a_j) = d(a_k, a_\ell) \quad \text{iff} \quad \{i,j\} = \{k,\ell\}. \tag{1}$$

We call this problem the one of finding a solution of the graceful immersion of K_N in Q_n.

We call a maximal subset $\{a_1, a_2, \ldots, a_N\}$ of vertices in Q_n satisfying (1) a graceful immersion of K_N in Q_n or a solution of the graceful immersion of K_N in Q_n.

Let $\{a_1, a_2, \ldots, a_N\}$ be such a solution, $a_i = (a_{i1}, a_{i2}, \ldots, a_{in})$, $(i = 1, 2, \ldots, N)$, $(a_{ij} = 0$ or $a_{ij} = 1)$, and

$$A = \begin{bmatrix} a_1 \\ a_2 \\ \vdots \\ a_N \end{bmatrix} = \begin{bmatrix} a_{11} & a_{12} & \cdots & a_{1n} \\ a_{21} & a_{22} & \cdots & a_{2n} \\ \vdots & \vdots & \vdots\vdots\vdots & \vdots \\ a_{N1} & a_{N2} & \cdots & a_{Nn} \end{bmatrix}$$

Then A is called a solution matrix of a graceful immersion of K_N in Q_n. The matrix A is the incidence matrix of a block design $B = \{B_1, B_2, \ldots, B_N\}$.

It is rather difficult to obtain the solution of a graceful immersion of K_N in Q_n, and, in general this is still an open problem. In this paper we give the solutions for N = 9,11,16. Thus we obtain three examples of block designs satisfying the conditions a) and b).

2. MAIN RESULT

LEMMA. For any N vertices in the hypercube Q_n, let K_N' denote the graph of pairwise distances determined by N vertices (i.e. the graph obtained by labelling each edge of the complete graph on those vertices with the corresponding distance in Q_n). Then the subgraph derived from the odd edges in K_N' is a cocycle of K_N'.

PROOF. Obviously, it is not possible that the subgraph derived from the odd edges is an odd cycle.

COROLLARY 1. If there are N vertices of Q_n which have pairwise different distances, then

a) $n \geq C_N^2$

b) If $n = C_N^2$, then there are two integers m_1 and m_2 such that

$$\begin{aligned} m_1 + m_2 &= N \\ m_1 \cdot m_2 &= \lceil n/2 \rceil \\ C_{m_1}^2 + C_{m_2}^2 &= \lfloor n/2 \rfloor \end{aligned} \tag{2}$$

G. Burosch, I. Havel and J.M. Laborde [3] proved the following theorem.

THEOREM. (2) $\Leftrightarrow$ $N = k^2$ or $k^2 + 2$ $(k = 1,2,\ldots)$

COROLLARY 2. If $n = C_N^2$, then a necessary condition for a graceful immersion of K_N in Q_n to exist is $N = k^2$ or $N = k^2 + 2$ for some k.

According to Corollary 2, we have the table in Figure 1.

k	0	1	2	3	4	5	...
$N = k^2$	0	1	4	9	16	25	...
$N = k^2+2$	2	3	6	11	18	27	...

Figure 1

By $n = C_N^2$, we obtain the table in Figure 2

N	0	1	2	3	4	6	9	11	16	...
n	0	0	1	3	6	15	36	55	120	...

Figure 2

The table in Figure 2 shows corresponding values for N and $n = C_N^2$. So when n and N are a pair of corresponding values in

Figure 2, we can try to look for a solution of a graceful immersion of K_N in Q_n.

We have obtained three solution-matrices of a graceful immersion of K_9, K_{11} and K_{16} in Q_{36}, Q_{55} and Q_{120}, respectively, as it follows.

a) A solution-matrix of a graceful immersion of K_9 in Q_{36} is

$$\begin{bmatrix}
000000000000000000000000000000000000 \\
100000000000000000000000000000000000 \\
111100000000000000000000000000000000 \\
111111111000000000000000000000001000 \\
111011111111111000000000100001111100 \\
100011111111111111111111100000000100 \\
100011111111111111111111111111110000 \\
111111111111111111111111111111111100 \\
111111111111111111111111111111111111
\end{bmatrix}$$

This solution-matrix is the incidence-matrix of the block design $B = \{B_1, B_2, \ldots, B_9\}$ on $s = \{x_1, x_2, \ldots, x_{36}\}$, where

$B_1 = \emptyset$
$B_2 = \{x_1\}$
$B_3 = \{x_1, x_2, x_3, x_4\}$
$B_4 = \{x_1, x_2, \ldots, x_9, x_{33}\}$
$B_5 = \{x_1, x_2, x_3, x_5, x_6, \ldots, x_{15}, x_{25}, x_{30}, x_{31}, \ldots, x_{34}\}$
$B_6 = \{x_1, x_5, x_6, \ldots, x_{25}, x_{34}\}$
$B_7 = \{x_1, x_5, x_6, \ldots, x_{32}\}$
$B_8 = \{x_1,, x_2, \ldots, x_{34}\}$
$B_9 = \{x_1, x_2, \ldots, x_{36}\}$

b) A solution-matrix of a graceful immersion of K_{11} in Q_{55} is

$$\begin{bmatrix}
0000000000000000000000000000000000000000000000000000000 \\
0110000000000000000000000000000000000000000000000000000 \\
0111110011111000000000000000000000000000000000000000000 \\
0110011111110111110000000000000000000000000000000000000 \\
0110010011110000011111111100000000000000000000000000000 \\
0001001011111111011111110000000000000000011111111111111 \\
0110010011110000011110000000000111111111111111111111111 \\
0000010011111111111111111111111111111111111111111111111 \\
0110001111111111111111111111111111111111111111111111111 \\
0111111111111111111111111111111111111111111111111111111 \\
1111111111111111111111111111111111111111111111111111111
\end{bmatrix}$$

This solution-matrix is the incidence-matrix of the block design $B = \{B_1, B_2, \ldots, B_{11}\}$ on $S = \{x_1, x_2, \ldots, x_{55}\}$, where B_i $(i = 1,2,\ldots,11)$ are determined similarly to the previous case.

c) A solution-matrix of a graceful immersion of K_{16} in Q_{120} is

```
000000000000000000000000000000000000000000000000000000000000000000000000000000000000000000000000000000000000000000000000
100000000000000000000000000000000000000000000000000000000000000000000000000000000000000000000000000000000000000000000000
100111000000000000000000000000000000000000000000000000000000000000000000000000000000000000000000000000000000000000000000
111110110000000000000000000000000000000000000000000000000000000000000000000000000000000000000000000000000000000000000000
111110100000011111111000000000000000000000000000000000000000000000000000000000000000000000000000000000000000000000000000
011111001100011111100011110000000000000000000000000000000000000000000000000000000000000000000000000000000000000000000000
000111110000011111100100011000000000111110111111111110000000000000000000000000000000000000000000000000000000000000000000
111110101110000011100011101111000000000101111111111010001111110000000000000000000000111000000000000000000000000000000000
111110101110000011100111010000000000111011111111000011100000011111111000000011111111110000000111100000000000000000000000
011111001100011100011000010001111111110111111110000001111110000000000000111111111111111111100000000000000000000000000000
100000010011100111100111110000000110000011111111110000000000100000000000011111111111111111111111111111111111111111111111
100111110000000001111000000001111110000001111111111111111111111111111111111111111111111111111111111111111111111111111111
000001110001100001111001110000000000111101111111111111111111111111111111111111111111111111111111111111111111111111111111
100000110000011111111111111111111111111111111111111111111111111111111111111111111111111111111111111111111111111111111111
100111111111111111111111111111111111111111111111111111111111111111111111111111111111111111111111111111111111111111111111
111111111111111111111111111111111111111111111111111111111111111111111111111111111111111111111111111111111111111111111111
```

This solution-matrix is the incidence-matrix of the block design $B = \{B_1, B_2, \ldots, B_{16}\}$ on $S = \{x_1, x_2, \ldots, x_{120}\}$, where B_i $(i = 1,2,\ldots,16)$ are determined similarly to case a).

REFERENCES

[1] P. Erdös, Chao Ko, R. Rado, Intersection theorems for systems of finite sets, Quart. J. Math. Oxford (2), 12(1961), 313-318.

[2] G.O.H. Katona, Extremal Problems for Hypergraphs, In Combinatorics 2, M. Hall, Jr., J.H. Lint eds. Mathematical Centre Tracts 56, Amsterdam 1974, pp. 13-42.

[3] G. Burosch, Ivan Havel, Jean-Marie Laborde, On distinct distance systems, Theorie de Graphes et Combinatoire, 3eme Colloque International, Marseille-Luminy, Juin 1986.

[4] Sun Shixin, Note sur l'immersion gracieuse de K_N dans Q_n, rapport de Recherche 637-M, IMAG, Janvier 1987.

[5] M. Mollard, C. Payan, Sun Shixin, Graceful Problems, Seventh Hungarian Colloquium on Finite and Infinite Combinatorics, Budapest, July 1987.

[6] Sun Shixin, A graceful immersion of K_9 in Q_{36}, J. Chengdu Inst. Radio Engineering. 1988; (2): 188-190.

[7] Kang Tai, Sun Shixin, Matrix method of graceful immersion of K_N in Q_n, J. Chengdu Teachers College. 1988; (1): 1-10.

[8] Sun Shixin, Kang Tai, Basic properties of solutions of graceful immersion of K_N in Q_n, J. Chengdu Teachers College. 1988; (1): 25-32.

One-Factorizations of Multigraphs

W. D. WALLIS, Southern Illinois University, Carbondale, Illinois

Standard graph-theoretic notions are assumed. All graphs are finite and undirected.

A *one-factor* in a graph G is a set of edges which between them contain every vertex of G precisely once each. A *one-factorization* is a set of one-factors which together contain each edge exactly once.

Clearly a graph must have an even number of vertices if it is to have a one-factor; to have a one-factorization it is necessary that the graph is regular (each vertex has the same valency). The complete graph K_{2n} satisfies these necessary conditions, and it is well-known that every such complete graph has a one-factorization (see, for example, [9, p. 439]). The complete bipartite graph $K_{n,n}$ has a one-factorization for every n - such factorizations are equivalent to Latin squares in an obvious way.

The complete multigraph λK_v has v vertices and there are λ edges joining each pair of vertices. One-factors and one-factorizations of λK_v are defined in the obvious way; if v is odd there can be no one-factor, and when v is even, say v = 2n, one can produce a one-factorization of λK_{2n} by taking λ copies of each factor in a one-factorization of K_{2n}.

Given a one-factorization of λK_{2n}, it may be that there exists an integer λ_1 (less than λ) such that some $\lambda_1(2n-1)$ of the one-factors form a one-factorization of $\lambda_1 K_{2n}$. In that case the one-factorization of λK_{2n} is called *decomposable*; otherwise it is *indecomposable*. When $\lambda > 1$, the one-factorizations of λK_{2n} just exhibited are all decomposable. It is natural to ask for which values of λ and n do there exist indecomposable one-factorizations of λK_{2n}.

A one-factorization is called *simple* if it contains no repeated one-factor. There is no direct correspondence between simplicity and indecomposability. However simple one-factorizations are useful in the discussion of indecomposable one-factorizations.

We start with two theorems which give infinite classes of indecomposable factorizations.

THEOREM 1 [4]. There is a simple indecomposable one-factorization of $(n-1)K_{2n}$ whenever 2n-1 is a prime.

PROOF. Suppose F is the one-factor

$$\infty 0,\quad 12,\quad 34,\quad \ldots,\quad (2n-3)(2n-2).$$

Let θ be a generator of the multiplicative group of non-zero integers modulo 2n-1. Then define F_{ij} to be derived from F by multiplying each entry by θ^i and then adding j: the edges in F_{ij} are

$$\infty j,\quad (\theta^i+j)(2\theta^i+j),\quad (3\theta^i+j)(4\theta^i+j),\quad \ldots$$

Then $\mathcal{F} = \{F_{ij} : 0 \le i \le n-2,\ 0 \le j \le n-2\}$ is a simple indecomposable one-factorization of $(n-1)K_{2n}$: simplicity is obvious, and indecomposability is left for the reader. □

THEOREM 2 [2]. Given $\lambda > 2$, let p be the smallest prime which does not divide λ. Then there exists an indecomposable one-factorization of $\lambda K_{2(\lambda+p)}$.

PROOF. We construct an indecomposable one-factorization $\mathcal{F}$ of the $\lambda K_{4(\lambda+p)}$ based on two disjoint vertex-sets L and R, each containing $\lambda+p$ vertices. We label the vertices of L as x_L, where x ranges through $Z_{\lambda+p}$, and similarly R is labelled with a subscript R. All arithmetic is carried out modulo $\lambda+p$.

For all i in $Z_{\lambda+p}$ other than i = 0, define

$$G_i = \{(a_L, a+i_R) : a \in Z_{\lambda+p}\} .$$

Each G_i is a one-factor of $\lambda K_{2(\lambda+p)}$. The first group of factors in $\mathcal{F}$ is the multiset consisting of G_1 taken $\lambda-p+1$ times, G_{1-p} taken $\lambda-1$ times, and each G_i, $3 \leq i \leq \lambda+p-1$, taken $i+1-p$ times. Each edge incident with both of the sets L and R occurs in exactly λ of the factors, except for those of the form $(i_L, i+1_R)$, $(i_L, i-p+1_R)$, $(i_L, i+2_R)$ and (i_L, i_R).

Next we describe factors F_i, $0 \leq i \leq \lambda+p-1$, which cover the remaining edges between L and R. F_0 consists of the $(i_L, i+1_R)$ for $0 \leq i \leq p-2$, $(p-1_L, 0_R)$, and all (j_L, j_R) for $p \leq j \leq \lambda+p-1$; and for $1 \leq i \leq \lambda+p-1$ define

$$F_i = \{(a+i_L, b+i_R) : (a_L, b_R) \in F_0\} .$$

We include all these F_i, $0 \leq i \leq \lambda+p-1$, in $\mathcal{F}$. By this stage $\mathcal{F}$ contains every edge from L to R exactly λ times, except that the edges $(i_L, i+2_R)$ are completely missing.

Now we construct the remaining factors. Notice that $\lambda+p$ is always odd: if λ is odd then p = 2 and if λ is even then p is odd. So both L and R contain an odd number of vertices. Select a near-one-factorization $\{H_1, H_2, \ldots, H_{\lambda+p}\}$ of the $K_{\lambda+p}$ based on L, and a near-one-factorization $\{K_1, K_2, \ldots, K_{\lambda+p}\}$ of the $K_{\lambda+p}$ based on R, with the property that for each i, i_R is the vertex missing from H_i and $i+2_R$ is the vertex missing from K_i. Then define

$$L_i = H_i \cup K_i \cup \{(i_L, i+2_R)\}.$$

$\mathcal{F}$ contains λ copies of each L_i, $0 \leq i \leq p+\lambda-1$. This completes the description of $\mathcal{F}$, which is clearly a one-factorization of $\lambda K_{2(\lambda+p)}$.

We must prove that this factorization $\mathcal{F}$ is indecomposable. For this we will only need to use the factors F_i, i = 0,..., $\lambda-p+1$. Suppose that we can partition the factors of $\mathcal{F}$ into two sets $\mathcal{F}_1$ and $\mathcal{F}_2$ such that each edge of the underlying complete graph appears λ_i times in $\mathcal{F}_i$, i = 1,2. Without loss of generality, suppose that F_0 lies in $\mathcal{F}_1$. This factor contains the edge (p_L,p_R), but does not contain $(p-1_L,p-1_R)$. Because these edges are to occur in $\mathcal{F}_1$ with the same frequency, $\mathcal{F}_1$ must have a factor which does not contain (p_L,p_R) but does contain $(p-1_L,p-1_R)$. But the only such factor is F_p. Thus $F_0 \in \mathcal{F}_1$ implies that $F_p \in \mathcal{F}_1$. Applying this argument again to F_p we prove that F_{2p} is in $\mathcal{F}_1$, and so on. Eventually we see that $F_i \in \mathcal{F}_1$ for $i = p,2p,\ldots,(\lambda+p)p$. But p and $\lambda+p$ have greatest common divisor 1, so we have shown that $F_0,F_1,\ldots,F_{\lambda+p-1}$ all belong to $\mathcal{F}_1$. These factors cover the edge $(0_L,0_R)$ λ times. It follow that $\mathcal{F}$ is indecomposable. □

The next three theorems give recursive constructions: given an indecomposable one-factorization of some λK_{2n}, we build another. It is convenient to set up some notation. We write N_k for the set of the first k positive integers. If $\mathcal{F}$ is a one-factorization of the K_{2n} based on vertex-set N_{2n}, and U is any ordered 2n-set, then $\mathcal{F}(U)$ is constructed by replacing i by the i-th member of U in every factor of $\mathcal{F}$, for every i. The factor derived from the factor F of $\mathcal{F}$ is denoted F(U). If $\mathcal{L}$ is a one-factorization of the $K_{n,n}$ based on the two vertex-sets N_n and $N_{2n}\backslash N_n$, and U and V are ordered n-sets, then $\mathcal{L}(U,V)$ is the one-factorization formed from $\mathcal{L}$ by the substitutions

$$(1, 2,\ldots, n) \to U$$
$$(n+1,n+2,\ldots,2n) \to V.$$

We say a one-factorization $\mathcal{F}$ of K_{2n} is *standardized* if the K_{2n} is based on N_{2n} and the i-th factor contains (i,2n). A one-factorization of $K_{n,n}$ is standardized if the vertex-sets are N_n and $N_{2n}\backslash N_n$ and the first factor is

$$\{(1,n+1),(2,n+2),\ldots,(n,2n)\}.$$

We now give three theorems which allow one to build cases of indecomposable factorizations from earlier cases.

THEOREM 3 [3]. Suppose there exists an indecomposable one-factorization of λK_{2n} for some $\lambda > 1$. Then there exists an indecomposable one-factorization of λK_{4n} which is not simple.

PROOF. Suppose $\mathcal{F} = \{F_1, F_2, \ldots, F_{\lambda(2n-1)}\}$ is an indecomposable one-factorization of the λK_{2n} based on N_{2n}. Select two ordered 2n-sets U and V, and a standardized one-factorization $\mathcal{L}$ of $K_{2n,2n}$. Then the factors in the one-factorization $\mathcal{L}(U,V)$, together with the $\lambda(2n-1)$ factors $F_i(U) \cup F_i(V)$, $1 \leq i \leq \lambda(2n-1)$, form a non-simple one-factorization of K_{4n}.

Suppose this factorization were decomposable: say $\{H_1, H_2, \ldots, H_s\}$ were a one-factorization of μK_{4n} where the H_i are among the factors listed. Write H_i' for the intersection of H_i with the K_{2n} based on U. Then $\{H_1', H_2', \ldots, H_s'\}$ is a one-factorization of the μK_{2n} based on U, and $\mathcal{F}(U)$ is indecomposable -- a contradiction. □

THEOREM 4 [3]. Suppose there exists an indecomposable one-factorization of λK_{2n}, for some $\lambda > 1$. Then there exists an indecomposable one-factorization of λK_{4n-2} which is not simple.

PROOF. Suppose $\mathcal{F}$ is an indecomposable one-factorization of λK_{2n}. Select two ordered 2n-sets $U = (u_1, u_2, \ldots, u_{2n})$ and $V = (v_1, v_2, \ldots, v_{2n})$ and write $U^* = U \backslash \{u_{2n}\}$, $V^* = V \backslash \{v_{2n}\}$. If the factor F of $\mathcal{F}$ contains the edge $\{i, 2n\}$ then F(U) contains $\{u_i, u_{2n}\}$; define F^* to be $F(U) \cup F(V)$ with $\{u_i, u_{2n}\}$ and $\{v_i, v_{2n}\}$ deleted and $\{u_i, v_i\}$ appended. Also select a standardized one-factorization $\mathcal{L}$ of $K_{2n-1,2n-1}$, and define $\mathcal{L}^*$ to consist of λ copies of the factorization $\mathcal{L}(U^*, V^*)$, with all λ copies of the factor

$$\left\{(u_1, v_1), (u_2, v_2), \ldots, (u_{2n-1}, v_{2n-1})\right\}$$

removed. Then

$$\mathcal{L}^* \cup \{F_1^*, F_2^*, \ldots, F_{\lambda(2n-1)}^*\}$$

is the required one-factorization of λK_{4n-2}. □

For the next result we need to know that the graph G(n,w), which is defined to have the integers modulo 2w has its vertices and edges xy whenever

$$w-n < x-y < w+n,$$

has a one-factorization whenever $n < w$. This has been proven by several authors ([6], [7], [8]). Following [7], we write P_d for the subgraph of $G(n,w)$ formed by the edges xy where $x-y \equiv k \pmod{2w}$. Clearly $G(n,w)$ equals the disjoint union

$$G(n,w) = P_{w-n+1} \cup P_{w-n+2} \cup \ldots \cup P_w.$$

P_w is a single one-factor. If we write δ for the greatest common divisor $(d,2w)$, then P_d consists of the δ cycles

$$i, d+i, 2d+i, \ldots, i-d \pmod{2w}$$

for $i = 0,1,\ldots,\delta-1$. When $2w/\delta$ is even, these cycles are even, so P_d has a one-factorization. In particular, P_d has a one-factorization whenever d is odd, and P_{w-1} has a one-factorization. Various possibilities exist according to the parities of w and d, but in every case $G(n,w)$ can be decomposed into some graphs which we know to have one-factorizations and the $P_{2x} \cup P_{2x+1}$ where $\delta = (2x,2w)$ is such that $2w/\delta$ is odd.

We now show (following [7]) that $P_{2x} \cup P_{2x+1}$ splits into four one-factors. We start with a decomposition $F_1 \cup F_2$ of P_{2x+1} into two one-factors:

$$F_1 = \left\{i,2x+i+1 \quad 4x+i+2,6x+i+3 \quad \ldots \quad i-4x-2,i-2x-1 \ : \ i \text{ odd}\right\}$$
$$\cup \left\{i-2x-1,i \quad 2x+i+1,4x+i+2 \quad \ldots \quad i-6x-4,i-4x-2 \ : \ i \text{ even}\right\}$$

while F_2 is obtained by exchanging the conditions "i odd" and "i even" in the above description. Two further factors, F_3 and F_4, are

$$F_3 = \left\{2x+i,4x+1 \quad 6x+1,8x+i \quad \ldots \quad -4x+i,-2x+1 \right.$$
$$4x+i+1,6x+i+1 \quad 8x+i+1,10x+i+1 \quad \ldots$$
$$\left. -2x+i+1,i+1 \quad i,2x+i+1 \ : \ i \text{ odd}\right\}$$

$$F_4 = \left\{i,2x+i \quad 4x+i,6x+i \quad \ldots \quad i-6x,i-4x \right.$$
$$2x+i+1,4x+i+1 \quad 6x+i+1,8x+i+1 \quad \ldots$$
$$\left. -4x+i+1,-2x+i+1 \quad i-x,i+1 \ : \ \text{odd}\right\}$$

These four factors do not form a one-factorization: the

edges in the set A,

$$A = \left\{ i,2x+1+i \quad i-2x,i+1 : i \text{ odd} \right\},$$

occur in F_1 and in $F_3 \cup F_4$, while the edges B,

$$B = \left\{ i,i-2x \quad i+1,2x+i+1 : i \text{ odd} \right\},$$

are omitted. But $(F_1 \backslash A) \cup B$ is a one-factor, and it together with F_2, F_3 and F_4 is the required one-factorization.

So we have

LEMMA 5. $G(n,w)$ always has a one-factorization. □

A nonempty set of edges of a one-factor F is called a *subfactor* of F. A one-factorization $\mathcal{F}$ of the λK_{2n} based on V is said to be *contained in* a one-factorization $\mathcal{G}$ of the λK_{2s} based on W if $V \subseteq W$ and if for each factor F in $\mathcal{F}$ there is a one-factor G in $\mathcal{G}$ such that F is a subfactor of G. This is also expressed by saying that $\mathcal{F}$ is *embedded in* $\mathcal{G}$.

THEOREM 6 [4]. Any indecomposable one-factorization of λK_{2n} can be embedded in a simple indecomposable one-factorization of λK_{2s} provided $\lambda \le 2n-1$ and $s \ge 2n$.

PROOF. Suppose $\mathcal{F} = \{F_{ij} : 1 \le i \le 2n-1,\ 1 \le j \le \lambda\}$ is a one-factorization of the K_{2n} with vertex-set $\{v_1,v_2,\ldots,v_{2n}\}$. Since there are at most λ copies of any given factor, and $\lambda \le 2n-1$, we can assume that $F_{ij} \ne F_{im}$ when $j \ne m$. We construct a one-factorization of the K_{2s} with vertex-set $V \cup Z_{2w}$, where $w \ge n$ and Z_{2w} is disjoint from V.

First take $H_1,H_2,\ldots,H_{2n-1}$ to be the factors in a one-factorization of $G(n,w)$. Write

$$K_{ij} = F_{ij} \cup H_i .$$

Then

$$\mathcal{K} = \{K_{ij} : 1 \le i \le 2n-1,\ 1 \le j \le \lambda\}$$

is a set of one-factors of the K_{2s}. Moreover they are all different: if K_{ij} equals $K_{\ell m}$ then it must be that $H_i = H_\ell$ and $F_{ij} = F_{\ell m}$; the former implies that $i = \ell$, but $F_{ij} = F_{im}$ implies

$i = m$.

We next need a sequence $(\{a_1,b_1\},\{a_2,b_2\},\ldots,\{a_{w-n},b_{w-n}\})$ of disjoint pairs of non-zero elements of Z_{2w}, such that $|a_r-b_r| = r$ for each r. This is easy: the pairs

$$\{1,w\},\{2,w-1\},\ldots,\{\lfloor w/2\rfloor,\lfloor (w+3)/2\rfloor\},$$
$$\{w+1,2w-1\},\{w+2,2w-2\},\ldots,\{\lfloor (3w-1)/2\rfloor,\lfloor (3w+2)/2\rfloor\}$$

have differences $1,2,\ldots,w-1$ once each; select the pairs with differences $1,2,\ldots,w-n$ and label them appropriately. Then write Y for the set of remaining elements of Z_{2w}: that is,

$$Y = Z_{2w}\setminus \bigcup_{r=1}^{w-n} \{a_r,b_r\},$$

and label the elements of Y as $\{y_1,y_2,\ldots,y_{2n}\}$. Now define

$$M_{ij} = \Big\{\{v_t,y_{t+j}+i\}: 1 \le t \le 2n\Big\} \cup \Big\{\{a_r+i,b_r+i\}: 1 \le r \le w-n\Big\}$$

(where the subscript on y is taken as an integer modulo 2n, for the purposes of addition). Then if

$$\mathcal{M} = \{M_{ij}: i \in Z_{2w},\ \ 1 \le j \le \lambda\},$$

$\mathcal{M}$ is a set of $2w\lambda$ distinct one-factors and $\mathcal{K} \cup \mathcal{M}$ is a simple one-factorization of λK_{2s} which contains the one-factorization $\mathcal{F}$.

Suppose $\mathcal{K} \cup \mathcal{M}$ is decomposable, and that certain of the factors constitute a factorization of μK_{2s}. Then the intersections of those factors with the K_{2n} based on V will be one-factors of the K_{2n} and will form a one-factorization of μK_{2n} as part of $\mathcal{F}$. Thus, if $\mathcal{F}$ is indecomposable, so is $\mathcal{K} \cup \mathcal{M}$. □

In addition to those already given, a number of other indecomposable factorizations have been constructed, mostly by computer.

THEOREM 7 [2,4]. There exist simple indecomposable one-factorizations of $2K_8$, $4K_8$, $2K_{10}$, $4K_{10}$, $5K_{10}$, $3K_{12}$, $4K_{12}$, $6K_{12}$, $8K_{12}$, $9K_{12}$, $3K_{14}$, $8K_{14}$, $9K_{14}$, $10K_{14}$, $5K_{16}$, $7K_{16}$, $8K_{16}$, $9K_{16}$, $10K_{16}$, $7K_{18}$, $8K_{18}$, $9K_{18}$, $10K_{18}$, $7K_{20}$, $8K_{20}$, $9K_{20}$, $10K_{20}$, $7K_{22}$, $10K_{22}$, $7K_{24}$, $10K_{24}$, $7K_{26}$ and $7K_{28}$, and indecomposable (but not simple) one-factorizations of $5K_8$, $6K_8$ and $12K_{16}$.

PROOF. The appendices give a construction for each factorization up to K_{20}. For the larger cases, see [4] and [2] In every case the K_{2n} is taken to have vertex-set $Z_{2n-1} \cup \{\infty\}$ and a list of initial blocks is given; each is to be developed like a starter. The case $6K_8$ is previously unpublished, and was constructed by J.H. Dinitz and the author. □

Observe that indecomposable (but not simple) one-factorizations of $2K_{10}$, $3K_{14}$ and $4K_{14}$ are given by Theorem 4.

It is easy to show that no indecomposable one-factorizations of $3K_6$ or $4K_6$ exist; we shall in fact prove a more general result in Theorem 11. Using this and the preceding results, we have:

THEOREM 8. An indecomposable one-factorization of λK_{2n} exists as follows:

$\lambda = 2$: if and only if $2n \geq 6$;
$\lambda = 3$: if and only if $2n \geq 8$;
$\lambda = 4$: if and only if $2n \geq 8$;
$\lambda = 5$: if $2n \geq 8$;
$\lambda = 6$: if $2n = 8$ or $2n \geq 12$;
$\lambda = 7$: if $2n \geq 16$;
$\lambda = 8$: if $2n \geq 12$;
$\lambda = 9$: if $2n \geq 12$;
$\lambda = 10$: if $2n \geq 14$;
$\lambda = 11$: if $2n = 24$ or $2n \geq 46$;
$\lambda = 12$: if $2n = 16$ or $2n \geq 30$.

Since there are exactly $1\cdot3\cdot\ldots\cdot(2n-1)$ one-factors of K_{2n}, the largest λ such that λK_{2n} has a simple factorization is

$$\lambda = 1\cdot3\cdot\ldots\cdot(2n-3),$$

and for a simple indecomposable factorization we must have

$$\lambda < 1\cdot3\cdot\ldots\cdot(2n-3).$$

However, this bound does not apply to indecomposable factorizations when simplicity is not required. We shall now derive a bound (which is probably very coarse) in that more general case.

By an *exact cover of depth* d on a set S we mean a collection of subsets of S, called blocks, such that each member of S belongs to exactly d blocks. (Repeated blocks are allowed.) If all the blocks are k-sets, the exact cover is called *regular* of degree k. An exact cover in S is *decomposable* if some proper subcollection of its blocks forms an exact cover on S. It is known (see [5]) that every sufficiently deep exact cover is decomposable: given s, there exists a positive integer D[s] such that any exact cover of depth greater than D[s] on an s-set is decomposable. It follows that there is also a maximum depth for a regular exact cover of degree k on an s-set: we denote it D[s,k].

LEMMA 9 [1]. Whenever $s \geq k \geq 1$,

$$D[s,k] < s^s \cdot \binom{sk+s+1}{s} . \qquad \square$$

THEOREM 10. If there is an indecomposable factorization of λK_{2n}, then

$$\lambda < [n(2n-1)]^{n(2n-1)} \binom{2n^3+n^2-n+1}{2n^2-n} .$$

PROOF. Suppose there is an indecomposable factorization $\mathcal{F}$ of λK_{2n}. Denote by S the set of all edges of K_{2n}: S is a set of size n(2n-1). The factors in $\mathcal{F}$, interpreted as subsets of S, form an n-regular exact cover of depth λ on S.
So

$$\lambda \leq D[n(2n-1),n],$$

giving the result. $\square$

As we said, this bound is very coarse. For example, in the case of λK_6, it is a little larger than 3×10^{31}. But we shall in fact show that no indecomposable one-factorization of λK_6 can exist for $\lambda \geq 3$. We assume that there is an indecomposable one-factorization $\mathcal{F}$ of λK_6 for some $\lambda \geq 3$ and derive a contradiction.

For notational convenience we assume K_6 to have vertices 0,1,2,3,4,5. Since the fifteen one-factors of K_6 form a one-factorization of $3K_6$, not all of them can appear in $\mathcal{F}$: say {01,23,45} is not represented. We denote the other possible one-factors as follows:

A = {01,24,35}	H = {03,15,24}
B = {01,25,34}	I = {04,12,35}
C = {02,13,45}	J = {04,13,25}
D = {02,14,35}	K = {04,15,23}
E = {02,15,34}	L = {05,12,34}
F = {03,12,45}	M = {05,13,24}
G = {03,14,25}	N = {05,14,23}

Denote the number of occurrences of A in $\mathcal{F}$ as a, and so on.

Since edge 01 must appear in λ factors, we have

$$a + b = \lambda. \tag{1}$$

One could derive fourteen more equations in this way. In particular, considering 23, 02, 14 and 34 we get

$$k + n = \lambda \tag{2}$$

$$c + d + e = \lambda \tag{3}$$

$$d + g + n = \lambda \tag{4}$$

$$b + e + \ell = \lambda \tag{5}$$

and (1) + (2) + (3) - (4) - (5) is

$$a + c - g + k - \ell = \lambda \tag{6}$$

We can assume that $a \geq \frac{1}{2}\lambda$ and $k \geq \frac{1}{2}\lambda$: if $a < \frac{1}{2}\lambda$ and $k < \frac{1}{2}\lambda$, then carry out the permutation (01)(45) on all members of $\mathcal{F}$ - it exchanges A with B and K with N and leaves {01,23,45} unchanged; if $a < \frac{1}{2}\lambda$ and $k \geq \frac{1}{2}\lambda$ then (01) is the relevant permutation; if $a \geq \frac{1}{2}\lambda$ and $k < \frac{1}{2}\lambda$ then use (45).

The factors {A,C,G,K,L} form a one-factorization of K_6, so a,c,g,k and ℓ cannot all be non-zero. The permutation (01)(23)(45) exchanges G and L, and leaves A, C and {01,23,45} unchanged; so without loss of generality we can assume $g \leq \ell$. Since a and k are positive, this means we can assume either c or g to be zero. But the equations derived from considering edges 03 and 45 are

$$f + g + h = \lambda, \tag{7}$$

$$c + f = \lambda, \tag{8}$$

whence $h = c-g$ and $c \geq g$. So $g = 0$. Substituting this into (6) and recalling that $a \geq \frac{1}{2}\lambda$ and $k \geq \frac{1}{2}\lambda$ we obtain $c-\ell \leq 0$. Counting occurrences of 12 we see that

$$f + i + \ell = \lambda;$$

from (8) we get $i = c-\ell$, and as i cannot be negative we have also: $c = \ell$ and $a = k = \frac{1}{2}\lambda$. Considering (04) we get $i + j + k = \lambda$; this and (1) and (2) now give $b = j = n = \frac{1}{2}\lambda$. Equation (5) tells us now that $e = a-c$, so $c \leq a \leq \frac{1}{2}\lambda$, and therefore from (8) f is non-zero. Since not all the members of the one-factorization $\{A,E,F,J,N\}$ can be represented, $e = 0$, whence c must equal $\frac{1}{2}\lambda$ also.

It is now easy to see that $e = g = i = m = 0$, and that the other ten factors each occur $\frac{1}{2}\lambda$ times. (The equations derived from edges 04 and 05 give the information about i and m.) If λ is odd, we have a contradiction. Otherwise we have $\frac{1}{2}\lambda$ duplicates of the one-factorization $\{A,B,C,D,F,H,J,K,L,N\}$ of $2K_6$, and $\mathcal{F}$ is decomposable. So we have proven the following Theorem.

THEOREM 11. There is no indecomposable one-factorization of λK_6 when $\lambda \geq 3$. □

REFERENCES

[1] N. Alon and K. A. Berman, Regular hypergraphs, Gordon's lemma, Steinitz' lemma and invariant theory, J. of Combinatorial Theory 43A(1986), 91-97.

[2] D. Archdeacon and J. H. Dinitz, Constructing indecomposable 1-factorizations of the complete multigraph. (to appear)

[3] A. H. Baartmans and W. D. Wallis, Indecomposable factorizations of multigraphs, Discrete Math. (to appear)

[4] C. J. Colbourn, M. J. Colbourn and A. Rosa, Indecomposable one-factorizations of the complete multigraph, J. of Austral. Math. Soc., 39A(1985), 334-343.

[5] J. E. Graver, A survey of the maximum depth problem for indecomposable exact covers, in Infinite and Finite Sets (Colloq. Math. Soc., J. Bolyai, North-Holland, Amsterdam, 1973), 731-743.

[6] A. Hartman, Tripling quadruple systems, Ars Combinatoria 10(1980), 255-309.

[7] R. G. Stanton and I. P. Goulden, Graph factorizations, general triple systems and cyclic triple systems, Aeq. Math. 22(1981), 1-28.

[8] G. Stern and H. Lenz, Steiner triple systems with given subspaces: Another proof of the Doyen-Wilson theorem, Boll. U. Mat. Ital. A (V), 17(1980), 109-114.

[9] A. P. Street and W. D. Wallis, Combinatorics: A first course, Charles Babbage Research Centre, Winnipeg, 1983.

APPENDIX I. SMALL SIMPLE FACTORIZATIONS

$2K_8$:	∞,0	1,6	2,3	4,5		
	∞,0	1,5	2,4	3,6		
$4K_8$:	∞,0	1,2	3,6	4,5		
	∞,0	1,4	2,3	5,6		
	∞,0	1,6	2,4	3,5		
	∞,0	1,5	2,4	3,6		
$2K_{10}$:	∞,0	1,4	2,6	3,7	5,8	
	∞,0	1,3	2,4	5,6	7,8	
$4K_{10}$:	∞,0	1,2	3,4	5,6	7,8	
	∞,0	1,4	2,7	3,5	6,8	
	∞,0	1,7	2,6	3,5	4,8	
	∞,0	1,3	2,6	4,7	5,8	
$5K_{10}$:	∞,0	1,2	3,4	5,6	7,8	
	∞,0	1,5	2,6	3,7	4,8	
	∞,0	1,4	2,3	5,7	6,8	
	∞,0	1,7	2,5	3,6	4,8	
	∞,0	1,7	2,4	3,5	6,8	
$3K_{12}$:	∞,0	1,4	2,7	3,10	5,8	6,9
	∞,0	1,7	2,3	4,5	6,10	8,9
	∞,0	1,7	2,9	3,5	4,6	8,10
$4K_{12}$:	∞,0	1,2	3,4	5,10	6,7	8,9
	∞,0	1,4	2,5	3,8	6,9	7,10
	∞,0	1,7	2,6	3,10	4,8	5,9
	∞,0	1,10	2,8	3,5	4,6	7,9
$6K_{12}$:	∞,0	1,2	3,4	5,6	7,9	8,10
	∞,0	1,3	2,4	5,6	7,8	9,10
	∞,0	1,4	2,5	3,6	7,9	8,10
	∞,0	1,4	2,8	3,6	5,9	7,10
	∞,0	1,5	2,9	3,7	4,8	6,10
	∞,0	1,6	2,7	3,8	4,9	5,10
$8K_{12}$:	∞,0	1,2	3,4	5,10	6,7	8,9
	∞,0	1,2	3,8	4,5	6,7	9,10
	∞,0	1,5	2,7	3,9	4,8	6,10
	∞,0	1,6	2,7	3,10	4,8	5,9
	∞,0	1,9	2,7	3,6	4,10	5,8
	∞,0	1,10	2,4	3,5	6,8	7,9
	∞,0	1,9	2,10	3,6	4,7	5,8
	∞,0	1,10	2,6	3,5	4,8	7,9

$9K_{12}$:	∞,0	1,4	2,10	3,5	6,8	7,9		
	∞,0	1,9	2,4	3,5	6,8	7,10		
	∞,0	1,2	3,6	4,7	5,8	9,10		
	∞,0	1,10	2,4	3,6	5,8	7,9		
	∞,0	1,2	3,4	5,6	7,8	9,10		
	∞,0	1,6	2,3	4,9	5,10	7,8		
	∞,0	1,6	2,7	3,8	4,9	5,10		
	∞,0	1,5	2,9	3,7	4,8	6,10		
	∞,0	1,7	2,6	3,10	4,8	5,9		
$3K_{14}$:	∞,0	1,12	2,3	4,5	6,7	8,10	9,11	
	∞,0	1,10	2,11	3,12	4,7	5,8	6,9	
	∞,0	1,7	2,10	3,8	4,9	5,11	6,12	
$8K_{14}$:	∞,0	1,11	2,3	4,5	6,7	8,9	10,12	
	∞,0	1,3	2,12	4,5	6,7	8,9	10,11	
	∞,0	1,4	2,6	3,7	5,8	9,11	10,12	
	∞,0	1,3	2,4	5,8	6,10	7,11	9,12	
	∞,0	1,11	2,12	3,5	4,8	6,10	7,9	
	∞,0	1,5	2,9	3,8	4,11	6,10	7,12	
	∞,0	1,9	2,7	3,8	4,12	5,10	6,11	
	∞,0	1,7	2,8	3,9	4,10	5,11	6,12	
$9K_{14}$:	∞,0	1,4	2,3	5,7	6,8	9,10	11,12	
	∞,0	1,2	3,6	4,5	7,9	8,10	11,12	
	∞,0	1,2	3,4	5,8	6,7	9,11	10,12	
	∞,0	1,11	2,4	3,6	5,8	7,9	10,12	
	∞,0	1,9	2,10	3,6	4,8	5,12	7,11	
	∞,0	1,9	2,11	3,7	4,10	5,8	6,12	
	∞,0	1,7	2,6	3,8	4,11	5,10	9,12	
	∞,0	1,6	2,10	3,9	4,8	5,12	7,11	
	∞,0	1,8	2,7	3,11	4,10	5,9	7,12	
$10K_{14}$:	∞,0	1,12	2,3	4,8	5,7	6,9	10,11	
	∞,0	1,12	2,6	3,5	4,7	8,9	10,11	
	∞,0	1,6	2,5	3,4	7,9	8,10	11,12	
	∞,0	1,2	3,8	4,7	5,6	9,11	10,12	
	∞,0	1,11	2,4	3,5	6,12	7,8	9,10	
	∞,0	1,11	2,12	3,10	4,7	5,8	6,9	
	∞,0	1,10	2,6	3,12	4,8	5,9	7,11	
	∞,0	1,9	2,7	3,8	4,12	5,10	6,11	
	∞,0	1,7	2,8	3,9	4,10	5,11	6,12	
	∞,0	1,8	2,11	3,7	4,9	5,10	6,12	
$5K_{16}$:	∞,0	1,2	3,11	4,5	6,12	7,8	9,10	13,14
	∞,0	1,7	2,9	3,5	4,6	8,10	11,13	12,14
	∞,0	1,13	2,14	3,12	4,11	5,8	6,9	7,10
	∞,0	1,12	2,6	3,10	4,8	5,14	7,11	9,13
	∞,0	1,9	2,12	3,8	4,14	5,10	6,11	7,13

$^7K_{16}$:	∞,0	1,2	3,4	5,6	7,8	9,10	11,12	13,14
	∞,0	1,14	2,4	3,5	6,8	7,9	10,12	11,13
	∞,0	1,5	2,14	3,6	4,7	8,11	9,12	10,13
	∞,0	1,4	2,13	3,14	5,9	6,10	7,11	8,12
	∞,0	1,6	2,11	3,8	4,13	5,10	7,12	9,14
	∞,0	1,10,	2,8	3,12	4,9	5,14	6,11	7,13
	∞,0	1,8	2,9	3,10	4,11	5,12	6,13	7,14

$^8K_{16}$:	∞,0	1,2	3,4	5,6	7,8	9,10	11,12	13,14
	∞,0	1,3	2,4	5,7	6,8	9,11	10,12	13,14
	∞,0	1,13	2,14	3,5	4,7	6,9	8,11	10,12
	∞,0	1,4	2,6	3,7	5,9	8,12	10,13	11,14
	∞,0	1,5	2,12	3,13	4,8	6,10	7,11	9,14
	∞,0	1,11	2,7	3,13	4,9	5,10	6,12	8,14
	∞,0	1,10	2,8	3,9	4,13	5,11	6,12	7,14
	∞,0	1,8	2,9	3,10	4,11	5,12	6,13	7,14

$^9K_{16}$:	∞,0	1,2	3,4	5,6	7,8	9,10	11,12	13,14
	∞,0	1,14	2,4	3,5	6,8	7,9	10,11	12,13
	∞,0	1,3	2,5	4,7	6,9	8,10	11,13	12,14
	∞,0	1,4	2,14	3,6	5,8	7,11	10,13	9,12
	∞,0	1,12	2,13	2,14	4,8	5,9	6,10	7,11
	∞,0	1,6	2,11	3,8	4,14	5,10	7,12	9,13
	∞,0	1,6	2,7	3,12	4,10	5,11	8,13	9,14
	∞,0	1,7	2,11	3,9	4,10	5,12	6,13	8,14
	∞,0	1,8	2,9	3,10	4,11	5,12	6,13	7,14

$^{10}K_{16}$:	∞,0	1,2	3,4	5,6	7,8	9,10	11,12	13,14
	∞,0	1,3	2,4	5,6	7,8	9,11	10,12	13,14
	∞,0	1,3	2,4	5,7	6,9	8,10	11,13	12,14
	∞,0	1,13	2,5	3,6	4,7	8,11	9,12	10,14
	∞,0	1,4	2,5	3,6	7,11	8,12	9,13	10,14
	∞,0	1,5	2,12	3,7	4,8	6,11	9,13	10,14
	∞,0	1,11	2,7	3,8	4,13	5,10	6,12	9,14
	∞,0	1,7	2,11	3,9	4,14	5,10	6,12	8,13
	∞,0	1,8	2,9	3,11	4,10	5,14	6,12	7,13
	∞,0	1,8	2,9	3,10	4,11	5,12	6,13	7,14

$^7K_{18}$:	∞,0	1,2	3,12	4,5	6,7	8,9	10,11	13,14	15,16
	∞,0	1,16	2,11	3,5	4,6	7,9	8,10	12,14	13,15
	∞,0	1,4	2,10	3,7	5,8	6,9	11,14	12,15	13,16
	∞,0	1,15	2,6	3,16	4,13	5,9	7,11	8,12	10,14
	∞,0	1,13	2,14	3,8	4,9	5,11	6,12	7,16	10,15
	∞,0	1,7	2,14	3,8	4,13	5,11	6,12	9,15	10,16
	∞,0	1,11	2,9	3,10	4,13	5,12	6,16	7,14	8,15

$^8K_{18}$:	∞,0	1,2	3,4	5,6	7,8	9,10	11,12	13,14	15,16
	∞,0	1,3	2,4	5,7	6,8	9,11	10,12	13,15	14,16
	∞,0	1,15	2,16	3,6	4,7	5,8	9,12	10,13	11,14
	∞,0	1,5	2,6	3,7	4,8	9,13	10,14	11,15	12,16
	∞,0	1,6	2,14	3,15	4,9	5,10	7,12	11,16	8,13
	∞,0	1,7	2,13	3,9	4,15	5,11	6,12	8,14	10,16
	∞,0	1,8	2,12	3,10	4,11	5,15	6,13	7,14	9,16
	∞,0	1,9	2,10	3,11	4,12	5,13	6,14	7,15	8,16

$^9K_{18}$:	∞,0	1,2	3,4	5,6	7,8	9,10	11,12	13,14	15,16
	∞,0	1,16	2,4	3,5	6,8	7,9	10,12	11,13	14,15
	∞,0	1,15	2,16	3,6	4,7	5,8	9,11	10,13	12,14
	∞,0	1,5	2,15	3,7	4,8	6,10	9,12	11,14	13,16
	∞,0	1,14	2,6	3,15	4,8	5,10	7,12	9,13	11,16
	∞,0	1,12	2,7	3,14	4,16	5,10	6,11	8,13	9,15
	∞,0	1,12	2,8	3,14	4,10	5,11	6,16	7,13	9,15
	∞,0	1,11	2,9	3,10	4,14	5,12	6,13	7,16	8,15
	∞,0	1,9	2,10	3,11	4,12	5,13	6,14	7,15	8,16

$^{10}K_{18}$:	∞,0	1,2	3,4	5,6	7,8	9,10	11,12	13,14	15,16
	∞,0	1,2	3,4	5,7	6,8	9,11	10,12	13,15	14,16
	∞,0	1,3	2,5	4,6	7,10	8,11	9,12	13,15	14,16
	∞,0	1,5	2,16	3,6	4,7	8,12	9,13	10,14	11,15
	∞,0	1,4	2,6	3,16	5,8	7,11	9,13	10,14	12,15
	∞,0	1,9	2,14	3,15	4,16	5,10	6,11	7,12	8,13
	∞,0	1,12	2,8	3,15	4,10	5,16	6,11	7,13	9,14
	∞,0	1,12	2,9	3,10	4,15	5,11	6,16	7,13	8,14
	∞,0	1,11	2,9	3,10	4,14	5,12	6,13	7,16	8,15
	∞,0	1,9	2,10	3,11	4,12	5,13	6,14	7,15	8,16

$^7K_{20}$:	∞,0	1,9	2,3	4,5	6,16	7,8	10,11	12,13	14,15	17,18
	∞,0	1,3	2,10	4,6	5,7	8,17	9,11	12,14	13,15	16,18
	∞,0	1,17	2,18	3,11	4,8	5,14	6,9	7,10	12,15	13,16
	∞,0	1,5	2,6	3,11	4,8	7,17	9,13	10,14	12,16	15,18
	∞,0	1,15	2,8	3,9	4,14	5,17	6,17	7,12	11,16	13,18
	∞,0	1,14	2,11	3,16	4,17	5,10	6,12	7,18	8,13	9,15
	∞,0	1,8	2,12	3,10	4,15	5,17	6,13	7,14	9,16	11,18

$^8K_{20}$:	∞,0	1,10	2,3	4,5	6,7	8,9	11,12	13,14	15,16	17,18
	∞,0	1,3	2,4	5,14	6,8	7,9	10,12	11,13	15,17	16,18
	∞,0	1,4	2,18	3,6	5,15	7,10	8,11	9,12	13,16	14,17
	∞,0	1,16	2,17	3,7	4,13	5,9	6,10	8,12	11,15	14,18
	∞,0	1,6	2,7	3,8	4,9	5,14	10,15	11,16	12,17	13,18
	∞,0	1,14	2,8	3,9	4,17	5,11	6,15	7,13	10,16	12,18
	∞,0	1,13	2,14	3,10	4,11	5,12	6,18	7,17	8,15	9,16
	∞,0	1,9	2,13	3,11	4,12	5,16	6,14	7,15	8,17	10,18

$9K_{20}$:	∞,0	1,2	3,4	5,6	7,8	9,10	11,12	13,14	15,16	17,18
	∞,0	1,18	2,4	3,5	6,8	7,9	10,12	11,13	14,16	15,17
	∞,0	1,4	2,5	3,6	7,10	8,11	9,12	13,16	14,17	15,18
	∞,0	1,5	2,17	3,7	4,8	6,10	9,13	11,15	12,16	14,18
	∞,0	1,15	2,16	3,17	4,18	5,10	6,11	7,12	8,13	9,14
	∞,0	1,14	2,8	3,16	4,10	5,18	6,12	7,13	9,15	11,17
	∞,0	1,8	2,9	3,15	4,16	5,12	6,13	7,14	10,17	11,18
	∞,0	1,12	2,10	3,11	4,15	5,13	6,14	7,18	8,16	9,17
	∞,0	1,10	2,11	3,12	4,13	5,14	6,15	7,16	8,17	9,18

$10K_{20}$:	∞,0	1,2	3,4	5,6	7,8	9,10	11,12	13,14	15,16	17,18
	∞,0	1,3	2,4	5,7	6,8	9,10	11,13	12,14	15,17	16,18
	∞,0	1,17	2,18	3,6	4,7	5,8	9,11	10,13	12,15	14,16
	∞,0	1,5	2,17	3,18	4,7	6,9	8,12	10,14	11,15	13,16
	∞,0	1,5	2,17	3,7	4,18	6,11	8,13	9,14	10,15	12,16
	∞,0	1,6	2,16	3,17	4,9	5,11	7,13	8,14	10,15	12,18
	∞,0	1,7	2,14	3,9	4,17	5,11	6,13	8,15	10,16	12,18
	∞,0	1,8	2,9	3,14	4,16	5,12	6,13	7,15	10,17	11,18
	∞,0	1,9	2,13	3,11	4,12	5,14	6,17	7,15	8,16	10,18
	∞,0	1,10	2,11	3,12	4,13	5,14	6,15	7,16	8,17	9,18

APPENDIX II. THREE NONSIMPLE FACTORIZATIONS

$5K_8$:	∞,0	1,2	3,4	5,6	
	∞,0	1,3	2,5	4,6	
	∞,0	1,5	2,6	3,4	twice
	∞,0	1,6	2,4	3,5	

$6K_8$:	∞,0	1,2	3,4	5,6	twice
	∞,0	1,4	2,5	3,6	
	∞,0	1,3	2,5	4,6	three times

$12K_{16}$:	∞,0	1,14	2,3	4,5	6,7	8,9	10,11	12,13	twice
	∞,0	1,4	2,7	3,14	5,13	6,12	8,10	9,11	five times
	∞,0	1,6	2,14	3,7	4,10	5,13	8,11	9,12	
	∞,0	1,10	2,5	3,14	4,9	6,13	7,11	8,12	
	∞,0	1,7	2,12	3,14	4,9	5,10	6,13	8,11	
	∞,0	1,11	2,8	3,14	4,10	5,12	6,9	7,13	
	∞,0	1,9	2,10	3,7	4,14	5,13	6,12	8,11	

Hamilton Cycles in Cartesian Products

W. D. WALLIS, Southern Illinois University, Carbondale, Illinois

Z. WANG, Suzhou Railway Teachers' College, Suzhou, China

1. INTRODUCTION

Kotzig has discussed in [2] the decomposition of cartesian products into Hamilton cycles and proved that the decomposition is possible if the two graphs can be decomposed into the same number of Hamilton cycles. However if one of them is nonhamiltonian, the question is still open.

Let G be a bridgeless cubic graph, not necessarily hamiltonian. It is the purpose of this paper to discuss the conditions for the cartesian product $G \times K_3$ to contain two Hamilton cycles with disjoint edges. When $G \times K_3$ contains two such Hamilton cycles, it has a one-factorization; then we can use the results in [4], [5] to draw some further conclusions on the one-factorizability of $G \times K_3$. This partially answers the question posed by Kotzig [3]: if G is a bridgeless cubic graph, does $G \times K_3$ necessarily have a one-factorization?

The graphs discussed in this paper are simple graphs, i.e., finite, loopless, undirected and without multiple edges (except that cycle graphs, defined below, may be multigraphs). The path with n edges is denoted P_n, and E is a graph consisting of a single edge. So a triangular prism $E \times K_3$ consists of two copies of K_3, together with three edges, one joining each vertex in one K_3 to the corresponding vertex in the other; these three edges are called *side edges*, and the others are *end edges*.

Cycle graphs were defined in [4]. Given a bridgeless cubic graph G, suppose it decomposes into a one-factor F_1 and a two-factor F_2. (Such a decomposition is always possible by Petersen's Theorem [1, p. 89].) Then the cycle graph G^* of G (with regard to the decomposition $F_1 \cup F_2$) is formed from G by contracting each cycle in F_2 to a point; a vertex of G^* is called an odd or even cycle point according as the cycle in F_2 from which it was formed had odd or even length. The reader should see [4] for more information on cycle graphs. When we use the notation G^*, it is assumed that the cycle graph came from the decomposition of a graph G into factors labeled F_1 and F_2.

2. SEVERAL LEMMAS

LEMMA 1. Suppose G has a subgraph H_1 isomorphic to $E \times K_3$ and G_1 is G with H_1 replaced by $H_2 = P_{2h+1} \times K_3$. If G has two edge disjoint Hamilton paths Q_1 and Q_2, while the three side edges of H_1 are all contained in Q_1 or Q_2 but not all in the same path, then G_1 also has two edge disjoint Hamilton paths Q_1' and Q_2'. If the end edges of H_2 and all edges not in H_2 are originally in Q_i, then they are contained in Q_i'. Q_i' has the same end points as Q_i does, $i = 1,2$. (Obviously $Q_1' \cup Q_2'$ contains all edges of inner triangles in H_2).

PROOF. In Figure 1 we suppose that the left side edge of H_1 is contained in Q_1 while the central one and right one are in Q_2. Here is a method to contract edge disjoint Hamilton paths Q_1' and Q_2' in G_1.

In Q_1 and Q_2 we replace the side edges of H_1 with paths in H_2 (in Figure 1 the side edges of H_1 and their corresponding paths in H_2 are of the same kind of line). All edges of inner

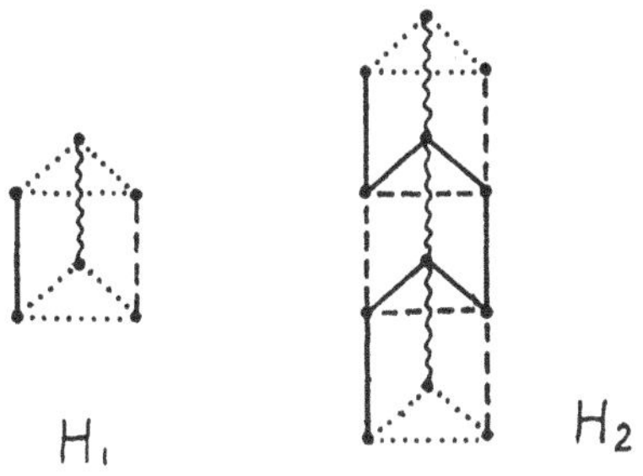

Figure 1

triangles in H_2 belong to Q_1 or Q_2 in the same way. If the side edges of H_1 belong to Q_i, then we put the corresponding side edges of the first, the third, the fifth, ..., copies of $E \times \overline{K}_3$ in H_2 into Q_i', $i = 1,2$: the left side edges into Q_1' and the central and the right into Q_2'. A similar classification can be done on the side edges of the second, the fourth, the sixth, ... copies of $E \times \overline{K}_3$: the right side edges go into Q_1' and the central and the left into Q_2' (see Figure 1). Obviously we can thus obtain two edge disjoint Hamilton paths Q_1' and Q_2'. □

Suppose x is a vertex of a graph G and u,v,w are the three vertices of K_3. In the cartesian product $G \times K_3$ we denote vertex x by u_x (v_x, w_x) according as x is in a copy of G relating with u (v,w). Thus vertices denoted by u (v,w) induce a copy of G while vertices with subscript x form a copy of K_3, $x \times K_3$.

LEMMA 2. Suppose x and y are two arbitrary vertices of an odd cycle C_{2k+1}. If s_1, s_2, t_1, t_2 are four distinct vertices of $C_{2k+1} \times K_3$, $s_1, s_2 \in \{u_x, v_x, w_x\}$, $t_1, t_2 \in \{u_y, v_y, w_y\}$, then $C_{2k+1} \times K_3$ has two edge disjoint Hamilton paths P_1 and P_2 such that s_i and t_i are the end points of P_i, $i = 1,2$. $P_1 \cup P_2$ does not contain s_1s_2 or t_1t_2.

PROOF. The vertices x and y divide C_{2k+1} into two x-y paths, one with odd length and the other with even length. For the case where the two paths have length 1 and 2 respectively, we construct in Figure 2(a) and (b) the required paths P_1 and P_2 in two cases respectively: (1) s_1, s_2, t_1, t_2 are in two copies of

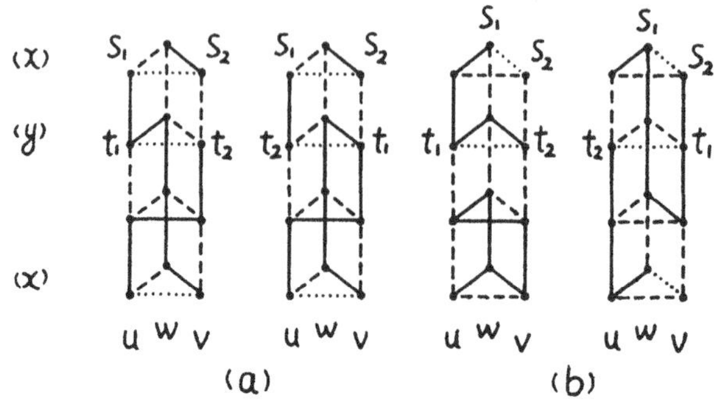

Figure 2

C_{2k+1} and (2) they are in three copies of C_{2k+1}. (For convenience of drawing the figure the $C_{2k+1} \times K_3$ has been cut in a certain copy of K_3, so the two triangular faces of each triangular prism in Figure 2 are actually the same copy of K_3. In the following figures this will also be done. P_1 and P_2 are shown in Figure 2, and subsequently, by dotted and solid lines respectively.)

By Lemma 1 this result can be extended to the case in which the two x-y paths, one with odd length and the other with even length, have arbitrary length. □

LEMMA 3. Suppose x and y are two arbitrary vertices of an even cycle C_{2k}. If s_1, s_2, t_1, t_2 are four distinct vertices of $C_{2k} \times K_3$, $s_1, s_2 \in \{u_x, v_x, w_x\}$, $t_1, t_2 \in \{u_y, v_y, w_y\}$, then $C_{2k} \times K_3$ has two edge disjoint Hamilton paths P_1 and P_2 such that s_i and t_i are two end points of P_i, $i = 1,2$. $P_1 \cup P_2$ does not contain s_1s_2 or t_1t_2.

PROOF. Again, the vertices x and y divide C_{2k} into two paths. We show in Figure 3 all cases in which both of the two paths have odd length or even length.

If both of the two paths have odd length we start on the assumption that one is of length 1 and the other of length 3. In Figure 3(a) we have constructed the required Hamilton paths P_1 and P_2 in four possible cases. By Lemma 1 the two x-y paths can be extended arbitrarily.

If the two x-y paths are both of even length, then two edge disjoint Hamilton paths can be constructed according to Figure

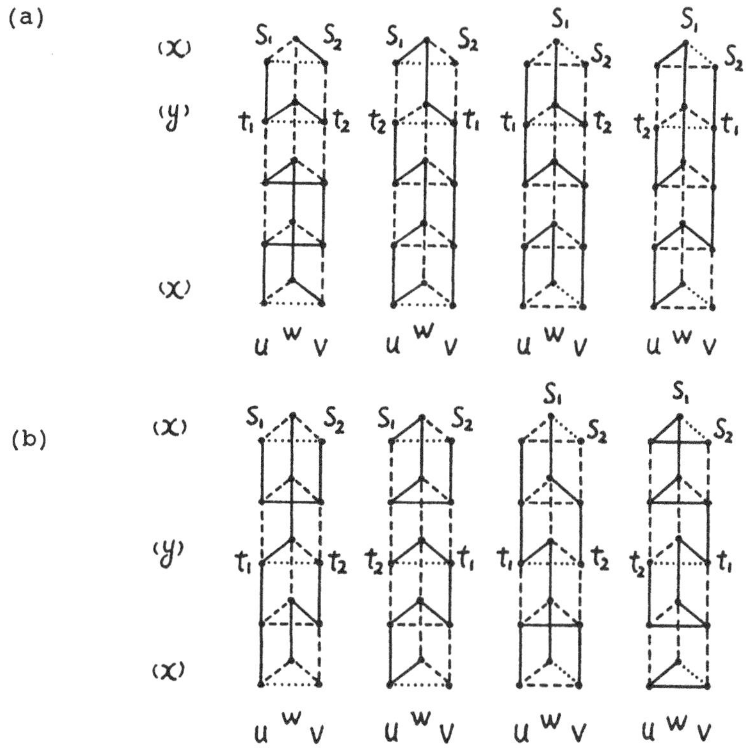

Figure 3

3(b). Once more the two paths can also be extended up to any even length. □

LEMMA 4. Suppose x and y are two arbitrary vertices of a cycle C. If s_1, s_2, t_1, t_2 are four distinct vertices of $C \times K_3$, $s_1, t_1 \in \{u_x, v_x, w_x\}$, $s_2, t_2 \in \{u_y, v_y, w_y\}$, then $C \times K_3$ has two edge disjoint Hamilton paths P_1 and P_2 such that s_i, t_i are the two end points of P_i, $i = 1,2$. $P_1 \cup P_2$ does not contain s_1t_1 or s_2t_2.

PROOF. First, suppose C is an odd cycle; x and y partition C into two paths, one wit odd length and the other with even length. By Lemma 1 we only need to prove the lemma for the case in which one path has length 1 and the other has length 2. In Figure 4(a) we have constructed two required edge disjoint Hamilton paths.

On the other hand, suppose C is an even cycle. C is divided by x and y into two paths of odd length or two paths of even length. s_1, s_2, t_1, t_2 can be located in two copies or three copies of C. Accordingly we construct in Figure 5(a) two edge disjoint Hamilton paths in four possible cases (two cases use each diagram). □

Obviously by joining s_i with t_i $(i = 1,2)$ in Lemma 4 we can obtain two edge disjoint Hamilton cycles which contain all edges of $C \times K_3$.

Suppose x,y are two arbitrary vertices of a graph G and $\bar{u},\bar{v} \in \{u,v,w\}$, $\bar{u} \neq \bar{v}$, $\bar{w} \in \{u,v,w\}\backslash\{\bar{u},\bar{v}\}$. In $G \times K_3$, $\bar{u}_x\bar{v}_x$ and $\bar{u}_y\bar{v}_y$ are called parallel edges, $\bar{u}_x\bar{v}_x$ and $\bar{u}_y\bar{w}_y$ (or $\bar{v}_y\bar{w}_y$) are called non-parallel edges.

Further results can be derived from Lemma 4 using Figure 4(b) and Figure 5(b). Figure 4(b) and Figure 5(b) are constructed by using the same method as in Lemma 1 and replacing $H_1 = E \times K_3$ with $H_2 = P_{2k+1} \times K_3$.

REMARK. On the assumption of Lemma 4, for a third arbitrary vertex z in the cycle C and for $\bar{u},\bar{v} \in \{u,v,w\}$, $\bar{u}_z\bar{v}_z$ is not parallel with both s_1t_1 and s_2t_2, and for the previously given $i_0 \in \{1,2\}$, we can always find P_1 and P_2 satisfying the requirement in Lemma 4 such that $\bar{u}_z\bar{v}_z$ is in P_{i_0}.

Obviously if x and y divide an even cycle into two even paths then we need only to point out that $\bar{u}_z\bar{v}_z$ is in P_{i_0} when z is in one of the two paths. In Figure 4(b) and Figure 5(b) we suppose $P_{i_0} = P_1$, $P_{i_0} = P_2$ respectively, and $\{\bar{u},\bar{v}\}$ is $\{u,v\}$ in each figure.

3. MAIN RESULTS

Suppose G_0 is a subgraph of a cycle graph G^* and U is the set of edges of F_1 which correspond to edges of G_0. The union of cycles in F_2 which correspond to vertices of G_0 plus the set

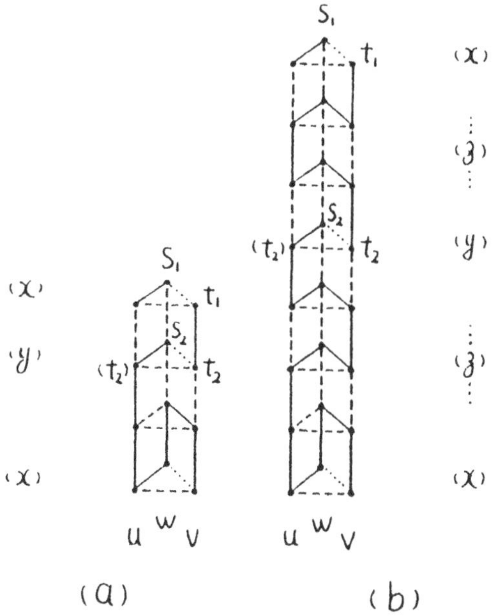

Figure 4

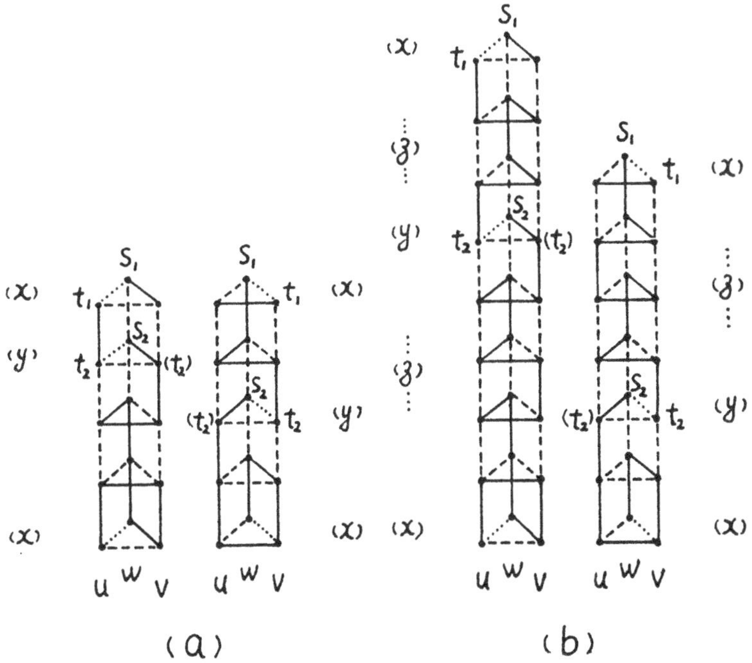

Figure 5

U produces a subgraph of G which we shall call a ***subgraph of G corresponding to*** G_0. The cartesian product of a subgraph of G corresponding to to G_0 with K_3 is called a subgraph of $G \times K_3$ corresponding to G_0.

THEOREM 1. If C^* is a cycle of a cycle graph G^*, then a subgraph of $G \times K_3$ corresponding to C^* contains two edge disjoint Hamilton cycles.

PROOF. Let $C^* = (g_1, g_2, \ldots, g_k)$, where each g_i corresponds to a cycle C_i of F_2, and edge $g_i g_{i+1}$ of C^* corresponds to edge e_i of F_1, where e_i joins vertex x_i of C_i to vertex y_{i+1} of C_{i+1}, for $i = 1, 2, \ldots, k$. Addition is carried out modulo k. Take $s_1 = u_{x_i}$, $t_1 = u_{y_i}$, $s_2 = v_{x_i}$, $t_2 = v_{y_i}$. By Lemma 2 and Lemma 3, $C_i \times K_3$ has two edge disjoint Hamilton paths P_{i_1} and P_{i_2} with end points s_1, t_1 and s_2, t_2 respectively, $i = 1, 2, \ldots, k$. Hence

$$Q_1 = \left[\bigcup_{i=1}^{k} P_{i_1}\right] + \{u_{x_i} u_{y_{i+1}} \mid i = 1, 2, \ldots, k;\ \text{subscripts modulo } k\}$$

$$Q_1 = \left[\bigcup_{i=1}^{k} P_{i_2}\right] + \{v_{x_i} v_{y_{i+1}} \mid i = 1, 2, \ldots, k;\ \text{subscripts modulo } k\}$$

give two edge disjoint Hamilton cycles in the subgraph of $G \times K_3$ corresponding to C^*. □

COROLLARY 1. If a cycle graph G^* contains a Hamilton cycle then $G \times K_3$ contains two edge disjoint Hamilton cycles. □

In fact, if G^* contains a Hamilton cycle then a subgraph of $G \times K_3$ corresponding to the Hamilton cycle should be a spanning subgraph of $G \times K_3$. Thus the conclusion follows.

Generally the order of a cycle graph G^* is much less than that of the graph G. Hence the above corollary is rather efficient in practical use. For some well-known nonhamiltonian graphs it can be easily seen that their cartesian products with K_3 all contain two edge disjoint Hamilton cycles. For example, consider the examples in Figure 6(a)-(g). (In each graph, a two-factor is presented by line like ——— or ---- while a one-factor is presented by ·····.) In each case a two-factor of each graph consists only of two edge disjoint cycles, so the cycle graph of each graph is so simple that it is just a multigraph of order 2 containing 2-cycles. By Corollary 1 the cartesian product of each graph with K_3 contains two edge disjoint Hamilton cycles.

DEFINITION 1. Suppose G_0 is a subgraph of a cycle graph G^*. If the subgraph of $G \times K_3$ corresponding to G_0 contains two edge disjoint Hamilton cycles, then G_0 is called a proper basic subgraph of G^*.

Obviously G^* always contains a proper basic subgraph. If subgraph G_0 of G^* is an isolated vertex or a cycle then G_0 is a proper basic subgraph of G^* by Lemma 4 and Theorem 1.

DEFINITION 2. Suppose G_0 is a subgraph of G^*. If a vertex g not contained in G_i is connected with G_0 by two edges e_1^* and e_2^* of G^*, $G_1 = G_0+g+\{e_1^*,e_2\}$, then G_1 is called a subgraph (of G^*) obtained by double-joining a vertex g to G_0 (through e_1^*,e_2^*), G_1 is also called a subgraph obtained by double-joining a vertex once to G_0.

THEOREM 2. By double-joining a vertex to a proper basic subgraph any (finite) number of times we always obtain a proper basic subgraph. In particular, if a spanning subgraph of G^* is obtained by double-joining a vertex to a proper basic subgraph a suitable number of times, then $G \times K_3$ contains two edge disjoint Hamilton cycles.

PROOF. If G_0 is a proper basic subgraph of a cycle graph G^*, then double-joining a vertex g_0 to G_0 produces a subgraph G_1 of G^*. We need only prove that G_1 is a proper basic subgraph.

Suppose g_0 is connected to vertices g_1,g_2 of G_0 by edges e_1^*

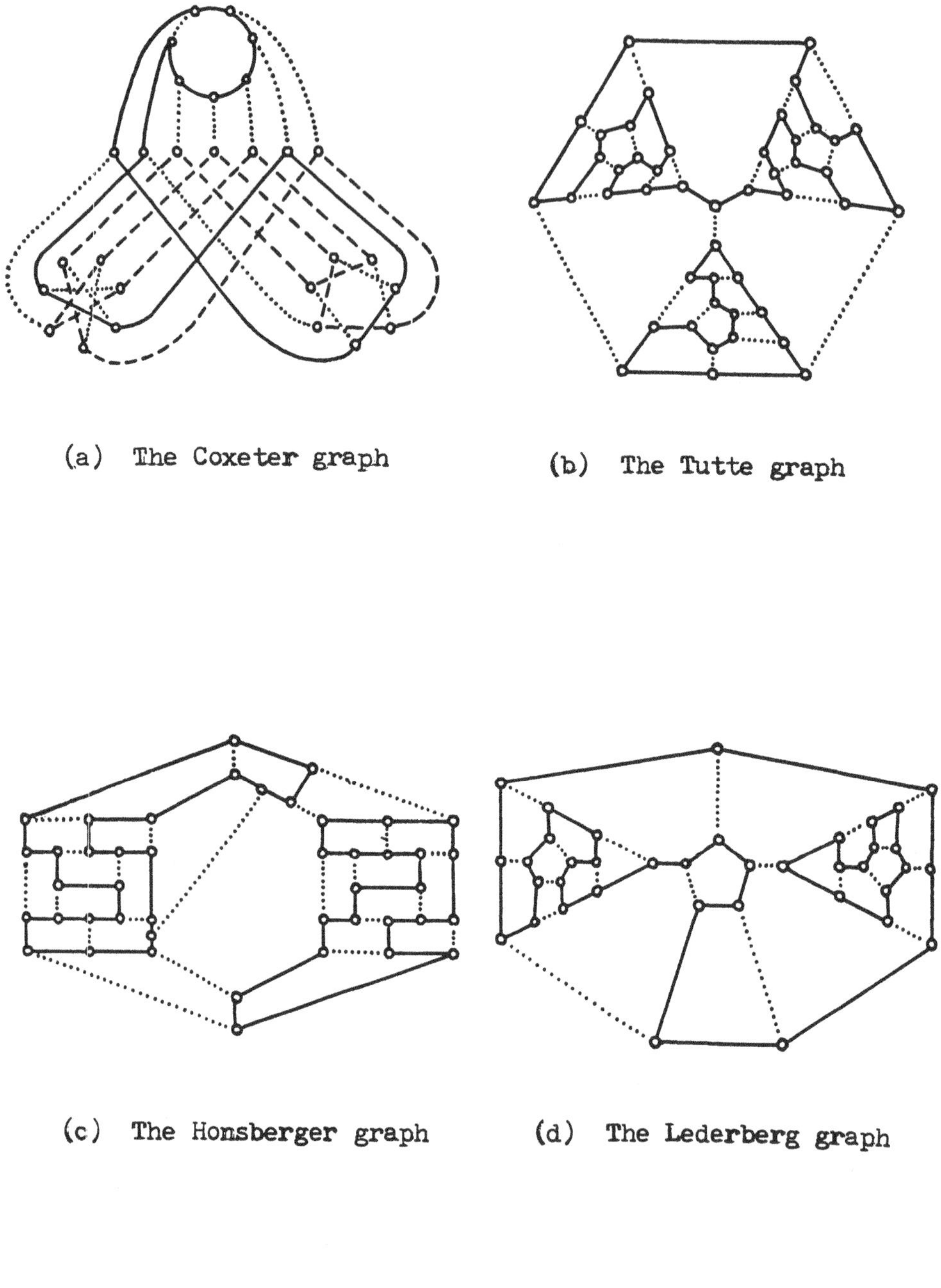

(a) The Coxeter graph

(b) The Tutte graph

(c) The Honsberger graph

(d) The Lederberg graph

Some famous non-hamilton cubic graphs
Figure 6

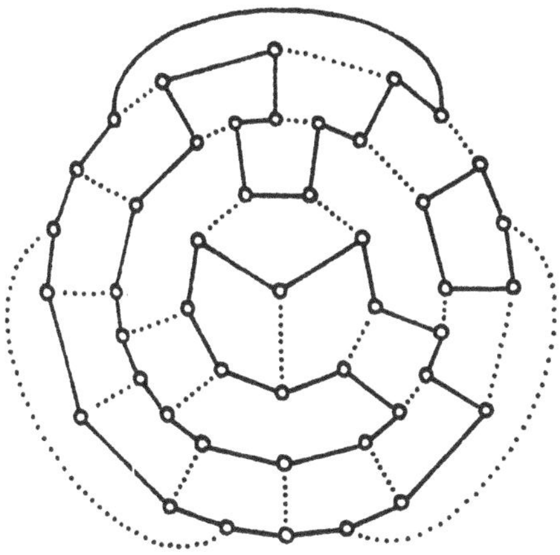

(e) The Grinberg graph

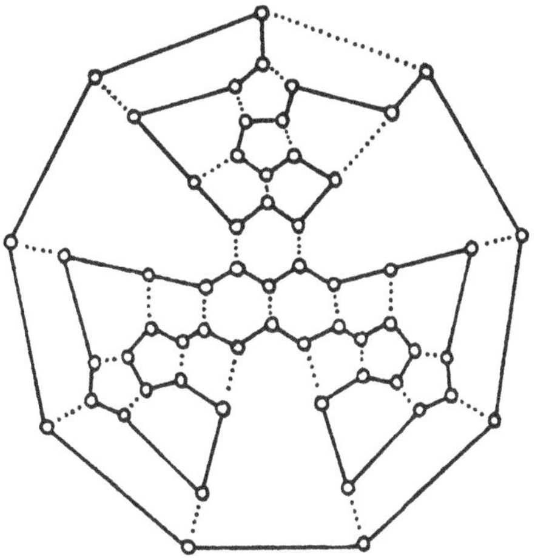

(f) The Hunter graph

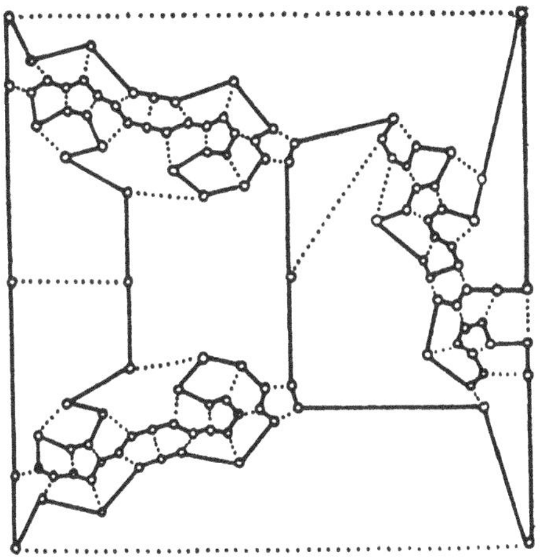

(g) The Walther graph

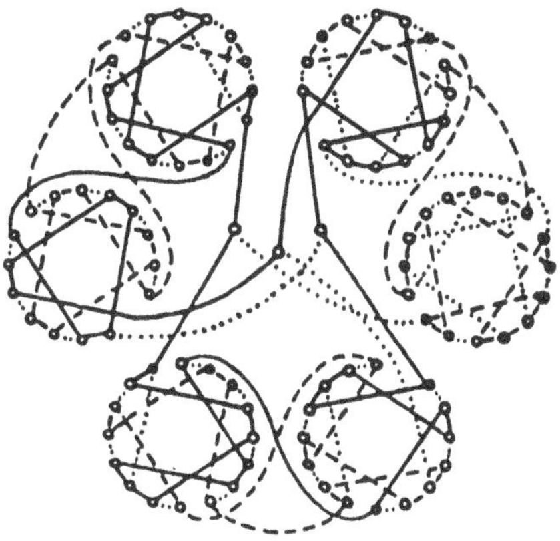

(h) The Horton graph

and e_2^* respectively. g_0, g_1 and g_2 correspond to cycles C_0, C_1 and C_2 in F_2 respectively while e_1^* and e_2^* correspond to edges e_1 and e_2 in F_1 respectively. e_1 joins a vertex x in C_0 with a vertex z in C_1. e_2 joins y in C_0 with r in C_2. Since G is cubic, $x \neq y$. By the definition of a proper basic subgraph, we know that a subgraph of $G \times K_3$ corresponding to G_0 contains two edge disjoint Hamilton cycles Q_1 and Q_2. $Q_1 \cup Q_2$ contains all edges of $z \times K_3$ and $r \times K_3$. Hence $z \times K_3$ contains some edge of Q_1, say a. Suppose e_{11} and e_{12} are edges of $e_1 \times \overline{K}_3$ which are adjacent to a and suppose that in $x \times K_3$ the end points of e_{11} and e_{12} are s_1, t_1 respectively. Similarly, $r \times K_3$ contains some edge of Q_2, say b. Suppose e_{21} and e_{22} are edges of $e_2 \times \overline{K}_3$ which are adjacent with b and suppose the end points in $y \times K_3$ of e_{21} and e_{22} are s_2 and t_2 respectively. By Lemma 4, $C_0 \times K_3$ contains two edge disjoint Hamilton paths P_1 and P_2 such that s_1,t_1 and s_2,t_2 are the end points of P_1 and P_2 respectively. Therefore the subgraph of $G \times K_3$ corresponding to G_1 contains two edge disjoint Hamilton cycles R_1 and R_2:

$$R_1 = (Q_1 - a + \{e_{11}, e_{12}\}) \cup P_1$$
$$R_2 = (Q_2 - b + \{e_{21}, e_{22}\}) \cup P_2. \qquad \square$$

COROLLARY 2. If an underlying simple graph of a cycle graph G^* is a tree then $G \times K_3$ contains two edge disjoint Hamilton cycles.

PROOF. Deleting some multiple edges of G^* such that every pair of adjacent vertices only have one edge joining them produces an underlying simple graph of G^*. Since G has no bridges, G^* has no bridges. That underlying simple graph is a tree means that every pair of adjacent vertices in G^* are joined by at least two edges. By Theorem 2 the conclusion follows. $\square$

For example, in every generalized Petersen graph P(n,k), the underlying simple graph of a cycle graph of P(n,k) is a star. By Corollary 2 we know that $P(n,k) \times K_3$ contains two edge disjoint Hamilton cycles. The graph in Figure 6(h) (its

2-factor consists of four disjoint cycles) has the cycle graph whose underlying simple graph is $K_{1,3}$. Hence the cartesian product of the Horton graph with K_3 contains two edge disjoint Hamilton cycles.

DEFINITION 3. Suppose G_0 is a subgraph of a cycle graph G^*. Two vertices g_1 and g_2 not contained in G_0 are joined to G_0 by edges e_1^* and e_2^* in G^* respectively. Say E_{12} is the set of multiple edges between g_1 and g_2 -- the set of all edges linking g_1 and g_2 which has at least two elements (sometimes they will be called multiple edges g_1g_2). Let $G_2 = (G_0)(g_1,g_2) + (E_{12} \cup \{e_1^*,e_2^*\})$. Then G_2 is called the graph *obtained by double-joining multiple edges* g_1g_2 to G_0 (through e_1^* and e_2^*).

THEOREM 3. Double-joining multiple edges to a proper basic subgraph of a cycle graph G^* produces a proper basic subgraph.

PROOF. Suppose G_0 is a proper basic subgraph of a cycle graph G^* and G_2 is produced by double-joining multiple edges g_1g_2 to G_0 through e_1^* and e_2^*. Let e_3^* and e_4^* be two arbitrary edges in multiple edges g_1g_2. e_1^* joins g_0 in G_0 with g_1 not in G_0 and e_2^* joins g_0' in G_0 with g_2 not in G_0. g_0, g_0', g_1, g_2 correspond to cycles C_0, C_0', C_1, C_2 of F_2 respectively and e_1^*, e_2^*, e_3^*, e_4^* correspond to edges e_1, e_2, e_3, e_4 of F_1 respectively. e_1 joins a vertex p in C_0 with a vertex x_1 in C_1, e_2 joins q in C_0' with y_2 in C_2, e_3 joins y_1 in C_1 with z_2 in C_2, e_4 joins z_1 in C_1 with x_2 in C_2, as shown in Figure 7 (C_0 and C_0' can be the same cycle).

Since G_0 is a proper basic subgraph, its corresponding subgraph of $G \times K_3$ contains two edge disjoint Hamilton cycles Q_1 and Q_2. Suppose $u_pv_p \in E(Q_1)$ and $\bar{u}_q\bar{v}_q \in E(Q_2)$, there $\{\bar{u},\bar{v}\} \subset \{u,v,w\}$. Select $\{\bar{\bar{u}},\bar{\bar{v}}\}$, $\{\hat{u},\hat{v}\} \subset \{u,v,w\}$ such that $\{\bar{\bar{u}},\bar{\bar{v}}\} \neq \{u,v\}$ or $\{\bar{u},\bar{v}\}$, $\{\hat{u},\hat{v}\} \neq \{\bar{u},\bar{v}\}$ or $\{\bar{\bar{u}},\bar{\bar{v}}\}$. By Lemma 4 and the following Remark we know that $C_2 \times K_3$ contains edge disjoint Hamilton

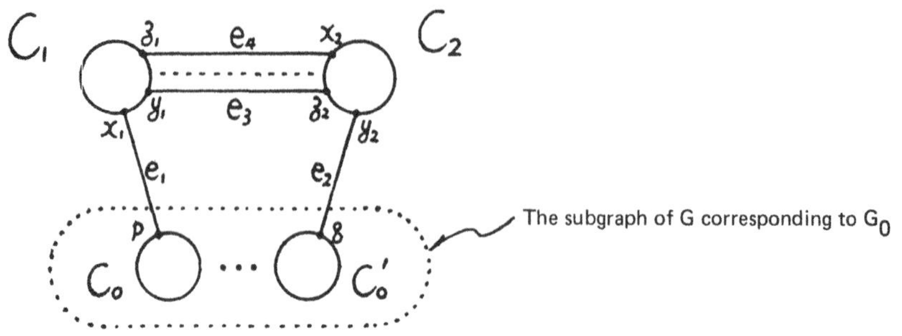

The subgraph of G corresponding to G_2
Figure 7

paths P_{21} and P_{22} such that their end points are $\bar{\bar{u}}_{x_2}, \bar{\bar{v}}_{x_2}$ and $\bar{u}_{y_2}, \bar{v}_{y_2}$ respectively when edge $\hat{u}_{z_2}\hat{v}_{z_2}$ of $z_2 \times K_3$ is in P_{22}. Similarly $C_1 \times K_3$ contains edge disjoint Hamilton paths P_{11} and P_{12} such that their end points are u_{x_1}, v_{x_1} and $\hat{u}_{y_1}, \hat{v}_{y_1}$ respectively and edge $\bar{\bar{u}}_{z_1}\bar{\bar{v}}_{z_1}$ is in P_{11}. Let

$$R_1 = (Q_1 \cup P_{11} \cup P_{21}) - \{u_p v_p, u_{z_1} v_{z_1}\} + \{u_p u_{x_1}, v_p v_{x_1}, \bar{\bar{u}}_{z_1}\bar{\bar{u}}_{x_2}, \bar{\bar{v}}_{z_1}\bar{\bar{v}}_{x_2}\},$$

$$R_2 = (Q_2 \cup P_{12} \cup P_{22}) - \{\bar{u}_q \bar{v}_q, \hat{u}_{z_2}\hat{v}_{z_2}\} + \{\bar{u}_q \bar{u}_{y_2}, \bar{v}_q \bar{v}_{y_2}, \hat{u}_{z_2}\hat{u}_{y_1}, \hat{v}_{z_2}\hat{v}_{y_1}\}.$$

Then R_1 and R_2 are two edge disjoint Hamilton cycles in the subgraph of $G \times K_3$ corresponding to G_2. Therefore G_2 is a proper basic subgraph of G^*. □

By Theorem 2 and Theorem 3 we know the following: if double-joining a vertex or double-joining multiple edges suitable times to a proper basic subgraph of a cycle graph G^* produces a spanning subgraph of G^*, then $G \times K_3$ contains two edge disjoint Hamilton cycles and hence has a one-factorization.

COROLLARY 3. In a cycle graph G^*, if there are at most two vertices not in a proper basic subgraph G_0 of G^*, then $G \times K_3$ has a one-factorization.

PROOF. If there is only one vertex g not in G_0, then double-joining the vertex g to G_0 produces a spanning subgraph of G^*, since G^* has no bridge. In this case the conclusion follows.

Suppose two vertices g_1 and g_2 are not in G_0. If it is possible for us to double-join one of $\{g_1,g_2\}$ to G_0, a spanning subgraph of G^* can be obtained by double-joining the vertex twice and the conclusion follows. Otherwise both g_1 and g_2 are joined to G_0 by one edge respectively. Then there are two cases as follows.

In one case there are multiple edges between g_1 and g_2. Then by Theorem 3 the conclusion follows. Otherwise g_1 and g_2 are joined by only one edge. Then g_1 and g_2 are even cycle points (A vertex of G^* is an even cycle point if and only if it has even degree in G^*). Subgraphs (disjoint) of $G \times K_3$ corresponding to G_0, g_1 and g_2 have four edge disjoint one-factors respectively (see [3]) and $G_0\}\{g_1,g_2\}$ is a spanning subgraph of G^*. Therefore $G \times K_3$ has four edge disjoint one-factors and has one-factorization. □

Kotzig has asked: if G is a basic subgraph, does $G \times K_3$ necessarily have a one-factorization? The authors [3] [4] have given several sufficient conditions for $G \times K_3$ to have a one-factorization and shown that if a basic subgraph G has a two-factor with at most three components then $G \times K_3$ has a one-factorization. In this paper the results obtained above (other than Theorem 1 and Corollary 1) can be used as new sufficient conditions for $G \times K_3$ to have a one-factorization. Using those theorems and corollaries we have the following.

COROLLARY 4. If a basic subgraph G has a two-factor with at most six components then $G \times K_3$ has a one-factorization.

PROOF. Suppose G is a connected graph. We construct all possible nonisomorphic simple connected graphs with at most six vertices (see, for example, the list in [1]) . If a cycle graph G^* has at most six vertices, then its underlying simple graph (also a connected graph) can be found in the list. Notice the fact: the end points of each bridge in an underlying simple graph of G^* are joined in G^* by at least two edges. In any possible case, given G^* we can always choose a cycle in its underlying simple graph as an initial proper basic subgraph of G^*. Then by double-joining to the cycle as many vertices as possible, using Theorem 2 and Corollary 3 yields the conclusion. □

By a detailed analysis of the results of this paper and [3] and [4] we can improve Corollary 4 to the case where G has a two-factor with at most seven components.

By using the same process as in the proof of Corollary 4 (but not using Corollary 3) in consideration of all (ten) connected simple graphs with at most four vertices we have the following.

COROLLARY 5. If a basic subgraph G has a two-factor with at most four components, then $G \times K_3$ contains two edge disjoint Hamilton cycles.

Hence we obtain another proof that the cartesian product of each graph in Figure 6 with K_3 has two edge disjoint Hamilton cycles.

REFERENCES

[1] F. Harary, Graph Theory, Addison-Wesley, Reading, Mass., 1969.

[2] A. Kotzig, Selected open problems in graph theory, Graph Theory and Related Topics, Academic Press, New York, 1979, 358-367.

[3] A. Kotzig, Problems and recent results on 1-factorization of cartesian products of graphs, Congressus Num. 21(1978), 457-460.

[4] W. D. Wallis and Wang Zhijian, On one-factorizations of cartesian products, Congressus Num. 49(1985), 237-245.

[5] W. D. Wallis and Wang Zhijian, Some further results on one-factorizations of cartesian products, J. Comb. Math. Comb. Comp., 1(1987), 221-234.

Nonexistence of Some Abelian Difference Sets

R. WEI, Department of Mathematics, Suzhou University, Suzhou, China

1. INTRODUCTION

A (v,k,λ) difference set in a finite group G of order v is a set D of k elements of G such that every element g of G, $g \neq e$, can be represented in precisely λ ways as $d_i d_j^{-1}$, where $d_i, d_j \in D$. A difference set D is called an abelian difference set when the group G is an abelian group.

Lander [5] lists 268 quadruples (v,k,λ,G) where G is an abelian group of order v and the parameters v, k, λ satisfy all the known necessary conditions for the existence of (v,k,λ) abelian difference sets with $k \leq v/2$, $k \leq 50$. Among them, 25 caess were not decided in [5]. Arasu [1] proved the nonexistence of two of those remaining cases. In this paper, we shall show the nonexistence of six other cases and therefore the undecided cases in Lander's table are reduced to 17. Precisely, we shall prove that there are no (704,38,2) abelian difference sets, no (27,13,6) abelian difference sets in $Z_3 \times Z_9$, and no (375,34,3) abelian difference sets in $Z_3 \times Z_5 \times Z_{25}$.

2. PRELIMINARIES

Suppose D is a difference set in an abelian group G (written in additive notation). Then an automorphism φ of G is called a multiplier if φ sends D onto D+g for some g in G. In particular, if the automorphism φ: $x \mapsto tx$ for some integer t is a multiplier t is also said to be a numerical multiplier. It is easy to see that all numerical multipliers of D form a group. The following well-known multiplier theorem is useful in proving the existence or nonexistence of abelian difference sets (see [3, 4]).

THEOREM 2.1. If G is an abelian group with a difference set D having parameters (v,k,λ) then, for any prime p with $p > \lambda$, $(p,v) = 1$ and p dividing $k-\lambda$, p is a multiplier.

Another theorem related to multipliers (see [6]) is also frequently used.

THEOREM 2.2. Let D be a (v,k,λ) difference set in an abelian group G. Then the group of numerical multipliers fixes at least one translate of D.

Using the above theorem, if a (v,k,λ) difference set D of an abelian group exists, we can assume without loss of generality that D is fixed by all of its numerical multipliers, since a translate of D is also a difference set with the same parameters.

Let B be a set of elements in a group G. Denote by ΔB the set of all $b_i b_j^{-1}$ with b_i, b_j in B. For other definitions and notations not given here see [2, 5].

3. MAIN RESULTS

The proofs in this paper are based on the following idea. Suppose α is a numerical multiplier of a (v,k,λ) difference set D in a group $G_1 \times G_2$ such that $\alpha h = h$ for every $h \in G_1$. Then the orbits of $G_1 \times G_2$ under α must have the form $\{h\} \times B$, where h is some element of G_1 and B is some orbit of G_2 under α. Since

D may be assumed to be fixed by α, we must have D = $\bigcup_{i=1}^{s}\{h_i\}\times B_i$, where $h_i \in G_1$ and B_i is some orbit of G_2 under α for i = 1,2, ...,s. So the nonexistence of some difference sets will follow from the properties of the orbits of G_2 under α. addition.

THEOREM 3.1. There does not exist any (704,38,2) abelian difference set.

PROOF. In Lander's table it is shown that (704,38,2) abelian difference sets do not exist except for undecided cases corresponding to the groups $(Z_4)^3\times Z_{11}$, $(Z_2)^2\times(Z_4)^2\times Z_{11}$, $(Z_2)^4\times Z_4\times Z_{11}$ and $(Z_2)^6\times Z_{11}$, where $Z_4\times Z_4\times Z_4$ is written briefly as $(Z_4)^3$ and so on. Let $G = G_1\times Z_{11}$ be any group from among the above four cases. Suppose D is a possible (704,38,2) difference set in G. By Theorem 1.1 and Theorem 1.2, 9 is a multiplier of D which fixes D. The congruence $9 \equiv 1 \pmod 8$ shows that every element h of G_1 satisfies 9h = h, so the orbits of G under the map $x \mapsto 9$ are

$$O_{h,1} = \{(h,0)\},$$
$$O_{h,2} = \{(h,1),\ (h,9),\ (h,4),\ (h,3),\ (h,5)\},$$
$$O_{h,3} = \{(h,2),\ (h,7),\ (h,8),\ (h,6),\ (h,10)\} \text{ for all } h \in G_1.$$

The set D must be a disjoint union of some of the above orbits. It is easy to check that for every nonzero element g of Z_{11} and any $h \in G_1$ (0,g) is contained in $\Delta O_{h,2}$ and $\Delta O_{h,3}$ twice each. Hence at most one orbit of the form $O_{h,2}$ and $O_{h,3}$ can be contained in D, otherwise it will contradict $\lambda = 2$. This shows that at least 33 elements of the 38 elements of D are contained in the orbits $O_{h,1}$ for $h \in G_1$. That is, at least 32 elements of D have the form (h,0) for some $h \in G_1\backslash\{0\}$. Now at least $33\cdot 32 = 1056$ elements of this form are contained in ΔD, since the difference of two such elements has the same form. But there are altogether 64 distinct elements in G which have the shape of (h,0). Therefore, at least one element of this form is contained in ΔD more than 16 times. This is a contradiction of the condition $\lambda = 2$, which means that a (704,38,2) abelian difference set in G cannot exist. The conclusion then follows.

In fact the nonexistence of (704,38,2) difference sets in $(Z_8)^2 \times Z_{11}$, $Z_2 \times Z_4 \times Z_8 \times Z_{11}$ and $(Z_2)^3 \times Z_8 \times Z_{11}$ has also been proved in the above proof, although which was proved in [5] in other ways.

PROPOSITION 3.2. there does not exist any (27,13,6) abelian difference set in $Z_3 \times Z_9$.

PROOF. Assume that there is a (27,13,6) difference set D in $Z_3 \times Z_9$ which is fixed by the multiplier 7. Then from the congruence $7 \equiv 1 \pmod 3$ we know that the orbits of $Z_3 \times Z_9$ under the map $x \mapsto 7x$ are

$$\begin{aligned} O_{h,1} &= \{(h,0)\}, \\ O_{h,2} &= \{(h,3)\}, \\ O_{h,3} &= \{(h,6)\}, \\ O_{h,4} &= \{(h,1),\ (h,7),\ (h,4)\}, \\ O_{h,5} &= \{(h,2),\ (h,5),\ (h,8)\}, \qquad \text{for all } h \in Z_3. \end{aligned}$$

As there are 9 elements in $\cup_{h\in Z_3} \cup_{i=1}^{3} O_{h,i}$, at least 4 of the 13 elements of D are contained in $\cup_{h\in Z_3}(O_{h,4} \cup O_{h,5})$. Since every $O_{h,4}$ and every $O_{h,5}$ contain 3 elements and D is a disjoint union of some orbits, at least 6 elements of D are contained in $\cup_{h\in Z_3}(O_{h,4} \cup O_{h,5})$. Notice that (0,3) is contained in $\Delta O_{h,4}$ and $\Delta O_{h,5}$ three times respectively and $\lambda = 6$, so exactly 6 elements of D are contained in $\cup_{h\in Z}(O_{h,4} \cup O_{h,5})$. Further, at most one element of D is contained in $\{(h,0),\ (h,3),\ (h,6)\}$ for any fixed $h \in Z_3$, otherwise $\lambda = 6$ will be contradicted. In other words, D contains at most 3 elements in $\cup_{h\in Z_3} \cup_{i=1}^{3} O_{h,i}$ and 6 elements in $\cup_{h\in Z_3}(O_{h,4} \cup O_{h,5})$. This contradicts $|D| = 13$ and the proof is complete.

Since (27,13,6) abelian difference sets do not exist in Z_{27} and H(27), the difference set of quadratic residues, is a (27,13,6) difference set in $Z_3 \times Z_3 \times Z_3$ by [5], we have the following.

THEOREM 3.3. The only (27,13,6) abelian difference sets are all in $Z_3 \times Z_3 \times Z_3$.

PROPOSITION 3.4. There does not exist any (375,34,3) difference set in $Z_3 \times Z_5 \times Z_{25}$.

PROOF. Suppose D is a (375,34,3) difference set in $G_1 \times Z_{25}$, where $G_1 = Z_3 \times Z_5$. Then 31 must be a multiplier of D which is fixed by the multiplier. By the congruences $31 \equiv 1 \pmod 3$ and $31 \equiv 1 \pmod 5$, we know that the orbits of $G_1 \times Z_{25}$ under the map $x \mapsto 31x$ are

$$
\begin{aligned}
O_{h,1} &= \{(h,1),\ (h,6),\ (h,11),\ (h,16),\ (h,21)\},\\
O_{h,2} &= \{(h,2),\ (h,12),\ (h,22),\ (h,7),\ (h,17)\},\\
O_{h,3} &= \{(h,3),\ (h,18),\ (h,8),\ (h,23),\ (h,13)\},\\
O_{h,4} &= \{(h,4),\ (h,24),\ (h,19),\ (h,14),\ (h,9)\},\\
O_{h,5} &= \{(h,0)\},\\
O_{h,6} &= \{(h,5)\},\\
O_{h,7} &= \{(h,10)\},\\
O_{h,8} &= \{(h,15)\},\\
O_{h,9} &= \{(h,20)\}, \qquad\qquad \text{for all } h \in G_1.
\end{aligned}
$$

Now $O_{h,1}$ cannot be contained in D since (0,5) appears four times in $\Delta O_{h,1}$. By the same reason, neither $O_{h,i}$ can be contained in D for $2 \leq i \leq 4$. But D can no longer be a difference set since (0,6) cannot appear in ΔD. This completes the proof.

ACKNOWLEDGEMENT

The author wishes to thank Professor L. Zhu for his helpful guidance and encouragement.

REMARK

Recently, we have learnt from Arasu [7] that the results of this paper are independently obtained by Arasu [8] using intersection numbers. We have also learnt that (64,28,12) difference sets in $Z_8 \times Z_8$ and in $Z_4 \times Z_{16}$ are found by Turyn [13], Davis [10,11], Dillon [12] and Arasu and Reis [9]. Moreover, the nonexistence of a (100,45,20) difference set in $Z_4 \times Z_5 \times Z_5$ is proved by McFarland [14]. Thus, there are now 14 cases in Lander's table left undecided.

REFERENCES

[1] K.T. Arasu, (81,16,3) abelian difference sets do not exist, J. Combinatorial Theory 43A(1986), 350-353.

[2] Th. Beth, D. Jungickel and H. Lenz, Design Theory, Bibliographisches Institut, Zurich, 1985.

[3] R.H. Bruck, Difference sets in a finite group, Trans. Amer. Math. Soc. 78(1955), 464-481.

[4] M. Hall, Jr. and H.J. Ryser, Cyclic incidence matrices, Canad. J. Math. 3(1951), 495-502.

[5] E.S. Lander, Symmetric Designs: An Algebraic Approach, London Math. Soc. Lecture Note Ser. Vol. 74, Cambridge U. P., Cambridge, 1983.

[6] R.L. McFarland and B.F. Rice, Translates and multipliers of abelian difference sets, Proc. Amer. Math. Soc. 68(1978), 375-379.

[7] K.T. Arasu, Recent result on difference sets, to appear in Proceedings of the IMA workshop on codes and designs in June 1988.

[8] K.T. Arasu, More missing entries in Lander's table could be filled, to appear in Arch. Math.

[9] K.T. Arasu and J. Reis, On abelian groups of order 64 that have difference sets, Wright State University Technical Report #1987.10 (1987).

[10] J. Davis, Difference sets in abelian 2-groups, Ph.D. Dissertation, U. of Virginia, 1987.

[11] J. Davis, Difference sets in abelian 2-groups, submitted.

[12] J.F. Dillon, Difference sets in 2-groups, to appear in Contemporary Math.: Finite Geometries and Designs, AMS 1989.

[13] R.J. Turyn, Personal communication.

[14] R.L. McFarland, Difference sets in abelian groups of order $4p^2$, submitted.

Extraneous Multipliers of Abelian Difference Sets

Wan-di WEI, Sichuan University, Chengdu, China

Shuhong GAO, Sichuan University, Chengdu, China

Qing XIANG, Sichuan University, Chengdu, China

1. INTRODUCTION

Let $(G,\cdot)$ be a group of order v. A k-subset D of G is called a (v,k,λ) difference set if the list of differences $d_1d_2^{-1}$, $d_1,d_2 \in G$, contains each non-identity element of G exactly λ times. The difference set D is said to be abelian (resp. cyclic) if G is abelian (resp. cyclic). The number $n = k-\lambda$ is called the order of the difference set. An automorphism α of G is called a multiplier of D if $D^{\alpha} = aDb$ for some $a,b \in G$. When a is the identity element, α is called a right multiplier. Let G be abelian. We know that for any positive integer t relatively prime to v, the mapping $x \mapsto x^t$ is an automorphism of G. If it happens to be a multiplier of D, then it is called a numerical multiplier of D. In this case we sometimes say that t is a multiplier by abuse of terminology. The following theorem suggests that the factors of n are an important source of multipliers of an abelian difference set of order n.

THEOREM A (R.H. Bruck [2]). Let D be a (v,k,λ) difference set in the abelian group G of order v and let m be a divisor of n. If $m > \lambda$ and m is relatively prime to v, and for any prime divisor p of m there exists an $f \in N \cup \{0\}$ with $t \equiv p^f$ (mod exponent of G), then the automorphism $\alpha_t: x \mapsto x^t$ is a numerical multiplier of D.

In fact, for all known abelian difference sets of order n, α_t is a multiplier for every positive integer t with $(t,n) = 1$ and $t|n$. It is, of course, not necessary that a multiplier t divides n. The multipliers not dividing n are called extraneous multipliers. H.B. Mann [6] and X.H. Wu [7] have proven that 2 (resp. 3) is a multiplier of a non-trivial cyclic difference set of order n only if $2|n$ (resp. $3|n$), and 5 is a multiplier of a non-trivial planar cyclic difference set only if $5|n$. In this paper we extend the above results of Mann and Wu to abelian difference sets and exclude -2, -3, and 4 from the possibility of being extraneous multipliers of planar abelian difference sets under certain conditions.

2. SOME KNOWN RESULTS

Let $(G,\cdot)$ be a gorup of order v, and $T = ZG$ the group ring of G over Z. Furthermore, let $D = \{d_1,d_2,\ldots,d_k\}$ be a k-subset of G $(0 < k < v)$. We put

$$T(D) = d_1 + d_2 + \cdots + d_k,$$

$$T(D^\alpha) = d_1^\alpha + d_2^\alpha + \cdots + d_k^\alpha, \text{ for any mapping } \alpha: G \to G, \text{ and}$$

$$S = \textstyle\sum_{g\in G} g,$$

$$\mathrm{supp}\left[\textstyle\sum_{g\in G} a_g g\right] = \{g \in G \mid a_g \neq 0\}.$$

Then D is a (v,k,λ) abelian group difference set in G if and only if

$$T(D)T(D^{-1}) = n+\lambda S$$

When $|D| = 1$ or $v-1$, D is called trivial. Let G be abelian and t an integer relatively prime to v. Then the automorphism $g \mapsto g^t$ of G is a multiplier if and only if $T(D^t) = cT(D)$ for some $c \in G$. The following theorems are important to our work.

THEOREM B (R.L. McFarland and B.F. Rice [3]). Let D be a (v,k,λ) difference set in an abelian group G. Then there is a translate of D which is fixed by all numerical multipliers of D.

THEOREM C (E.C. Johnsen [4]). Let D be a non-trivial (v,k,λ) abelian difference set in G, and -1 be a multiplier of D. Then

1) v and λ are even

2) any positive integer t relatively prime to v is a multiplier of D.

3. MAIN RESULTS

A theorem proved in [7] can be easily extended in the following form.

THEOREM 1. A prime p is an extraneous multiplier of a non-trivial (v,k,λ) abelian difference set D if and only if p satisfies either of the following congruence equations

$$T(D)^{p-1} \equiv c \pmod{p} \tag{1}$$

$$T(D)^{p-1} \equiv c-S \pmod{p} \tag{2}$$

for some $c \in G$.

The proof is similar to that in [7], so it is omitted.

COROLLARY 1. 2 is never an extraneous multiplier of a non-trivial (v,k,λ) abelian difference set D.

PROOF. If 2 is an extraneous multiplier of D, then

$$T(D) \equiv c \pmod{2}$$

or

$$T(D) \equiv c-S \pmod{2}.$$

Either of them is impossible as D is non-trivial. This completes the proof.

COROLLARY 2. 5 is never an extraneous multiplier of a non-trivial planar abelian difference set, but may be an extraneous multiplier of some non-trivial cyclic difference set with $\lambda > 1$.

PROOF. By Theorem C, -1 is never a multiplier of a non-trivial planar abelian difference set. Thus the proof in [7] can be easily extended to the present situation. Hence the former assertion of the corollary follows. For the latter assertion we give the following example.

EXAMPLE. Let $G = Z_{11}$. Then $D = \{1,3,4,5,9\}$ is an (11,5,2) cyclic difference set, and admits 5 as its extraneous multiplier because $5 \equiv 3^3 \pmod{11}$.

LEMMA 1. t is a multiplier of a (v,k,λ) difference set D in an abelian group G if and only if t is a multiplier of $\bar{D}$ where $\bar{D}$ is the complement of D in G.

PROOF. Let D be a (v,k,λ) abelian difference set. Then $\bar{D}$ is also an abelian difference set. If t is a multiplier of D, then $T(D^t) = cT(D)$ for some $c \in G$. So

$$\begin{aligned} T(\bar{D}^t) &= T(G^t) - T(D^t) \\ &= S - cT(D) \\ &= c(S-T(D)) = cT(\bar{D}). \end{aligned}$$

Therefore t is a multiplier of $\bar{D}$. By the symmetry of D and $\bar{D}$, the converse is also true. This completes the proof.

THEOREM 2. If 2 does not divide $\gcd(v,\lambda)$, then 3 is never an extraneous multiplier of a non-trivial (v,k,λ) abelian difference set D. If 2 divides $\gcd(v,\lambda)$, and if 3 is an extraneous multiplier of a non-trivial abelian difference set D, then

1) If $3|k$, then $n \equiv 1 \pmod 3$, $\lambda \equiv 2 \pmod 6$, $v \equiv 4 \pmod 6$, and -1 is a multiplier of D.

2) If $3 \nmid k$, then $n \equiv 1 \pmod 3$, $\lambda \equiv 0 \pmod 6$, and -1 is a

multiplier of D.

PROOF. By Lemma 1 and the basic relation $\lambda(v-1) = k(k-1)$, we may assume $2k < v$ without any loss of generality. Suppose that 3 is an extraneous multiplier of D. By Theorem B, there is a translate of D fixed by 3. We may assume that D itself is fixed by 3. By Theorem 1 we have

$$T(D)^2 \equiv 1 \pmod 3, \text{ if } 3|k$$

or

$$T(D)^2 \equiv 1-S \pmod 3, \text{ if } 3\nmid k.$$

That is

$$nT(D) + \lambda kS \equiv T(D^{-1}) \pmod 3, \text{ if } 3|k \tag{3}$$

or

$$nT(D) \equiv T(D^{-1}) \pmod 3, \text{ if } 3|k. \tag{4}$$

Assume that $2\nmid \gcd(v,\lambda)$. If (4) holds, then $n \equiv 1 \pmod 3$ and

$$T(D) \equiv T(D^{-1}) \pmod 3$$

i.e.

$$T(D) = T(D^{-1}).$$

Hence -1 is a multiplier of D. By Theorem C, v and λ are even, a contradiction.

Now suppose (3) holds. When $3|\lambda$, (3) is reduced to (4) which has just been dealt with. So we assume $3\nmid\lambda$. From (3), we have

$$nT(D) - T(D^{-1}) \equiv -\lambda kS \pmod 3 \text{ and } 3|k,$$

so

$$v = |\text{Supp}(nT(D) - T(D^{-1}))| \leq 2k$$

contradicting the assumption $2k < v$. This proves the first part of the theorem.

Assume $2|\gcd(v,\lambda)$. If (4) holds, -1 is still a multiplier of D. Therefore v,λ are even and $n \equiv 1 \pmod 3$, $\lambda \equiv 2 \pmod 6$ and $v \equiv 4 \pmod 6$. If (3) holds, an analogous argument to that in the proof of the first part can be used to prove $3|\lambda$. So (3) is reduced to $T(D) \equiv T(D^{-1}) \pmod 3$ i.e. $T(D) = T(D^{-1})$. Therefore, -1 is a multiplier of D. $n \equiv 1 \pmod 3$ and $\lambda \equiv 0$

(mod 6). This completes the proof.

EXAMPLES: When $2|\gcd(v,\lambda)$ for each of the two cases in Theorem 2, we have a (v,k,λ) abelian difference set D with 3 as an extraneous multiplier,

1) Let $G = Z_4 \times (Z_2)^2$. It is known [5] that there is a (16,6,2) difference set D in G with -1 as a multiplier. By Theorem C, 3 is an extraneous multiplier of D.

2) Let $G = (Z_5)^3 \times (Z_2)^5$. It is known [5] that there is a (4000,775,150) abelian difference set D with multiplier -1. So 3 is an extraneous multiplier of D.

THEOREM 3. -2 is a multiplier of a (v,k,λ) planar abelian difference set only if the group G is of type $(3)^\alpha$, an elementary abelian group.

PROOF. Let D be a (v,k,λ) planar abelian difference set in G with exponent v. If -2 is a multiplier of D, then for $d \in D$, we have $-2d \in D$, $4d \in D$ and $d-(-2d) = 3d = 4d-d$. Therefore $\lambda = 1$ implies $3d \equiv 0 \pmod{v^*}$ for any $d \in D$. Let $d_1, d_2 \in D$ and $d_1-d_2 = 1$. Then $3d_1-3d_2 \equiv 0 \pmod{v^*}$ and then $v^* = 3$ and $G = (Z_3)^\alpha$. This completes the proof.

THEOREM 4. If $\gcd(2,n) = 1$, then 4 is an extraneous multiplier of a planar abelian difference set of order n only if $G = (Z_3)^\alpha$.

PROOF. Let D be an abelian planar difference set in G. Since $\lambda = 1$, it follows that all terms d_id_j are distinct for $i \neq j$. Thus $T(D)^2 - T(D^2) = 2\ \sum_{i<j} d_id_j$ has exactly $k(k-1)/2$ terms. We put $T(D^2)-T(D^2) = 2P(D)$.

Suppose 4 is a multiplier of D. By Theorem B, we may assume

$$T(D^4) = T(D).$$

Then

$$T(D)^4 \equiv T(D) \pmod 2.$$

Multiplying both sides of this equation by $T(D^{-1})^2$, we have

$$T(D)^2(n+S)^2 \equiv (n+S)T(D^{-1}) \pmod 2$$

then

$$(T(D^2)+2P(D))(n+S)^2 \equiv (n+S)T(D^{-1}) \pmod 2$$

i.e.

$$T(D^2)(n+S) \equiv (n+S)T(D^{-1}) \pmod 2$$

Since $2\nmid n$, we have

$$T(D^2) \equiv T(D^{-1}) \pmod 2$$

so

$$T(D^2) = T(D^{-1}).$$

This means that -2 is a multiplier of D and then $G = (Z_3)^{\alpha}$. The proof is thus complete.

THEOREM 5. -3 is never a multiplier of a non-trivial (v,k,λ) planar abelian difference set D when $v > 7$.

PROOF. If $3|n$, then 3 is a multiplier of D by Theorem A. If -3 is also a multiplier of D, then -1 is a multiplier of D. That is impossible.

Now suppose that $3\nmid n$ and -3 is a multiplier of D. We may assume that

$$T(D^{-3}) = T(D)$$

Thus

$$T(D^{-1})^3 \equiv T(D) \pmod 3$$

Replacing D^{-1} by D in the above equation, we obtain

$$T(D^3) \equiv T(D^{-1}) \pmod 3. \tag{5}$$

Multiplying both sides of (5) by $T(D^{-1})$, we have

$$T(D)^2(n+S) \equiv T(D^{-1})^2 \pmod 3. \tag{6}$$

If $n \equiv 1 \pmod 3$, then $v = n^2+n+1 \equiv 0 \pmod 3$, so $(v,v-3) \neq 1$, contradicting the fact that -3 is a multiplier of D.

When $n \equiv -1 \pmod 3$ (6) becomes

$$T(D)^2 + T(D^{-1})^2 \equiv 0 \pmod 3$$

i.e.

$$T(D^2) + T(D^{-2}) \equiv P(D) + P(D^{-1}) \pmod 3.$$

Then we find that

$$\mathrm{Supp}T(D^2) \cap \mathrm{Supp}T(D^{-2}) = \mathrm{Supp}P(D) \cap \mathrm{Supp}P(D^{-1})$$

$$\mathrm{Supp}T(D^2) \cup \mathrm{Supp}T(D^{-2}) = \mathrm{Supp}P(D) \cup \mathrm{Supp}P(D^{-1}).$$

Since

$$|\mathrm{Supp}T(D^2)| = |\mathrm{Supp}T(D^{-2})| = k$$

$$|\mathrm{Supp}P(D)| = |\mathrm{Supp}P(D^{-1})| = k(k-1)/2$$

we have

$$2k = 2k(k-1)/2$$

i.e.

$$k = 3, \; n = 2$$

also contradicting the assumption $v > 7$. This completes the proof.

Note that when $v \lesssim 7$, the conclusion is no longer true. In fact, for $G = Z_7$, $\{1,2,4\}$ is a $(7,3,1)$ difference set in G and -3 is a multiplier of it by the fact that $-3 \equiv 2^2 \pmod 7$.

ACKNOWLEDGEMENT

The authors wish to thank Professor Benfu Yang for helpful discussions.

REFERENCES

[1] L.D. Baumert, Cyclic Difference Sets, Lecture Notes in Math. 182, Springer-Verlag, 1972.

[2] R.H. Bruck, Difference sets in a finite group, Trans. Amer. Math. Soc. 78(1955).

[3] R.L. Mcfarland and B.F. Rice, Translates and multipliers of abelian difference sets, Proc. Amer. Math. Soc. 68(1978), 375-379.

[4] E.C. Johnsen, The inverse multiplier for abelian group difference sets, Canad. J. Math. 16(1964), 787-796.

[5] E.S. Lander, Symmetric Designs, An Algebraic Approach, Cambridge U. P., 1983.

[6] H.B. Mann, Some theorems on difference sets, Canad. J. Math. 4(1952), 464-481.

[7] X.H. Wu, Extraneous multiplier of cyclic difference sets, J. Combinatorial Theory 42A(1986), 259-269.

Large Sets of KTS (V)

WU Lisheng, Department of Mathematics, Suzhou University, Suzhou, China

A Kirkman Triple System on v points is denoted KTS(v).

DEFINITION 1. A large set of Kirkman Triple System of order v (LKTS(v)) is a set of v-2 pairwise disjoint KTS(v) on some v-set V.

DEFINITION 2. A transitive Kirkman Triple System of order v (TKTS(v)) is a KTS(v) which has a sharply transitive automorphism group.

LEMMA 1 ([1, Problem 1]). If there exists a TKTS(v) and an LKTS(v), then there exists an LKTS(3v).

LEMMA 2 ([1, Problem 3]). If there exists a TKTS(v), then there exists a TKTS(3v).

LEMMA 3 ([2]). There exists an LKTS(v) when v = 33,51,105.

It is well known that TKTS(3v) exists for each prime power $v \equiv 1 \pmod 6$. There are no examples of TKTS(3v) for $v \equiv 5 \pmod 6$, $v > 5$. Our Lemma 4 will provide some of them.

LEMMA 4. There exists a TKTS(3v) when v = 11,17,35.

PROOF. Let $Z_{6t+5} = \{0,1,\ldots,6t+4\}$,

$$V = \{x_0,x_1,\ldots,x_{6t+4}\} \cup \{y_0,y_1,\ldots,y_{6t+4}\} \cup \{z_0,z_1,\ldots,z_{6t+4}\}$$
$$\alpha = (x_0,x_1,\ldots,x_{6t+4})(y_0,y_1,\ldots,y_{6t+4})(z_0,z_1,\ldots,z_{6t+4})$$
$$\beta = (x_0,y_0,z_0)(x_1,y_1,z_1)\ \cdots\ (x_{6t+4},y_{6t+4},z_{6t+4})$$

where α and β are permutations on V.

Consider the following basic triples:

$$\Delta_0 = (x_0,y_0,z_0), \qquad \Delta_l = (x_{p_l},x_{q_l},x_{r_l}), \quad 1 \le l \le t$$
$$\Delta_{t+1} = (x_a,x_b,x_c) \qquad \Delta_{t+2} = (x_d,x_e,x_f),$$
$$\Delta_k = (x_{u_k},y_{v_k},z_{w_k}), \quad t+3 \le k \le 3t+2$$

Write

$$\mathbb{B} = \{\alpha^i\Delta_0:\ 0 \le i \le 6t+4\} \cup \{\alpha^i\beta^j\Delta_s:\ 0\le i\le 6t+4,\ 0\le j\le 2,\ 1\le s\le 3t+2\}.$$

We can easily check that $(V,\mathbb{B})$ is a STS(3(6t+5)), if

(1) $\{|q_l-p_l|,\ |r_l-q_l|,\ |p_l-r_l|:\ 1 \le l \le t\} \cup \{|a-b|,\ |d-e|\} = \{1,2,\ldots,3t+2\}$

(2) $\{u_k-v_k,\ v_k-w_k,\ w_k-u_k:\ t+3\le k\le 3t+2\} \cup \{c-a,c-b,f-d,f-e\} = Z_{6t+5}-\{0\}$.

Furthermore, if

(3) there exists a $\tilde{\mathbb{B}}$ with $\{\beta^j\Delta_k:\ 1 \le k \le t+2,\ 0 \le j \le 2\} \subset \tilde{\mathbb{B}} \subset \{\Delta_0\} \cup \{\beta^j\Delta_s:\ 1 \le s \le 3t+2,\ 0 \le j \le 2\}$, $|\tilde{\mathbb{B}}| = 6t+5$, and $i_B \in Z_{6t+5}$ for each $B \in \tilde{\mathbb{B}}$ such that $\cup\{\alpha^{i_B}B:\ B \in \tilde{\mathbb{B}}\} = V$, and denote

$$\{\Delta_0\} \cup \{\beta^j\Delta_s:\ 1 \le s \le 3t+2,\ 0 \le j \le 2\} - \tilde{\mathbb{B}} = \{B_m:\ 6t+5 \le m \le 9t+6\},$$

then $(V,\mathbb{B})$ will be a KTS(3(6t+5)) with parallel classes

$$\{\mathbb{B}_m:\ 0 \le m \le 9t+6\}:\quad \mathbb{B}_0 = \{\alpha^{i_B}B:\ B \in \tilde{\mathbb{B}}\},$$

$$\mathbb{B}_m = \begin{cases} \alpha^m B_0, & \text{if } 1 \le m \le 6t+4 \\ \{\alpha^i B_m : 0 \le i \le 6t+4\}, & \text{if } 6t+5 \le m \le 9t+6 \end{cases}$$

as $G = \{\alpha^i \mathbb{B}^j : 0 \le i \le 6t+4,\ 0 \le j \le 2\} \subset \mathrm{Aut}((V,\mathbb{B}))$ is sharply transitive on V, we get a TKTS(3(6t+5)).

We cannot construct TKTS(3(6t+5)) for all t but here are some concrete examples: we only give the Δ_s and $\mathbb{B}_0$.

t=1:

$\Delta_0 = (x_0,y_0,z_0)$, $\Delta_1 = (x_0,x_1,x_3)$, $\Delta_2 = (x_0,x_4,y_3)$,
$\Delta_3 = (x_0,x_5,y_7)$, $\Delta_4 = (x_0,y_9,z_3)$, $\Delta_5 = (x_0,y_4,z_5)$,

$$\mathbb{B}_0 = \{\alpha^6\Delta_1, \alpha^6\beta\Delta_1, \alpha^{10}\beta^2\Delta_1, \Delta_2, \beta\Delta_2, \alpha^5\beta^2\Delta_2, \alpha^5\Delta_3, \alpha^8\beta\Delta_3, \alpha^7\beta^2\Delta_3, \alpha^{10}\beta\Delta_4, \alpha\Delta_5\}$$

t=2:

$\Delta_0 = (x_0,y_0,z_0)$, $\Delta_1 = (x_0,x_1,x_4)$, $\Delta_2 = (x_0,x_5,x_7)$,
$\Delta_3 = (x_0,x_6,y_3)$, $\Delta_4 = (x_0,x_8,y_4)$, $\Delta_5 = (x_0,y_{16},z_{11})$,
$\Delta_6 = (x_0,y_1,z_{10})$, $\Delta_7 = (x_0,y_2,z_7)$, $\Delta_8 = (x_0,y_{15},z_6)$,

$$\mathbb{B}_0 = \{\alpha^3\Delta_0, \alpha^{10}(\Delta_1,\beta\Delta_1,\beta^2\Delta_1), \alpha^{16}(\Delta_2,\beta\Delta_2,\beta^2\Delta_2), \alpha^{12}(\Delta_3,\beta\Delta_3,\beta^2\Delta_3), \alpha^5(\Delta_4,\beta\Delta_4,\beta^2\Delta_4), \alpha^8\Delta_5, \alpha^7\Delta_6, \Delta_7, \alpha^2\Delta_8\}.$$

t=5

$\Delta_0=(x_0,y_0,z_0)$, $\Delta_1=(x_0,x_1,x_3)$, $\Delta_2 = (x_0,x_4,x_9)$,
$\Delta_3=(x_0,x_6,x_{13})$, $\Delta_4=(x_0,x_{11},x_{21})$, $\Delta_5=(x_0,x_{12},x_{20})$,
$\Delta_6=(x_0,x_{16},y_8)$, $\Delta_7=(x_0,x_{18},y_9)$, $\Delta_8=(x_0,y_{15},z_3)$,
$\Delta_9=(x_0,y_1,z_{11})$, $\Delta_{10}=(x_0,y_{11},z_7)$, $\Delta_{11}=(x_0,y_7,z_1)$,
$\Delta_{12}=(x_0,y_6,z_{10})$, $\Delta_{13}=(x_0,y_2,z_{14})$, $\Delta_{14}=(x_0,y_3,z_{16})$,
$\Delta_{15}=(x_0,y_5,z_{19})$, $\Delta_{16}=(x_0,y_{18},z_5)$, $\Delta_{17}=(x_0,y_{17},z_2)$,

$$\mathbb{B}_0 = \{\alpha^2\Delta_0, \alpha^{20}(\Delta_1,\beta\Delta_1,\beta^2\Delta_1), \alpha^{29}(\Delta_2,\beta\Delta_2,\beta^2\Delta_2), \alpha^{13}(\Delta_3,\beta\Delta_3,\beta^2\Delta_3), \alpha^6(\Delta_4,\beta\Delta_4,\beta^2\Delta_4), \alpha^{10}(\Delta_5,\beta\Delta_5,\beta^2\Delta_5), \alpha^{24}(\Delta_6,\beta\Delta_6,\beta^2\Delta_6), \alpha^{25}(\Delta_7,\beta\Delta_7,\beta^2\Delta_7), \alpha^2(\Delta_{13},\beta\Delta_{13},\beta^2\Delta_{13}), \alpha^{15}(\Delta_{14},\beta\Delta_{14},\beta^2\Delta_{14}), \alpha^9(\Delta_{15},\beta\Delta_{15},\beta^2\Delta_{15}), \Delta_9, \beta^2\Delta_{10}, \beta\Delta_{11}, \alpha\beta^2\Delta_{12}\}.$$

Combining Lemma 1 through Lemma 4, we get the following:

THEOREM. There exists an LKTS($3^n m$) when $n \geq 1$, $m = 11,17,35$.

REFERENCES

[1] R.H.F. Denniston, Further cases of double resolvability, J. Combinatorial Theory A 26(1979), 298-303.

[2] A. Rosa, Intersection properties of Steiner systems, Ann. Discrete Math. 7(1980), 115-128.

Hadamard Matrices

Ming-yuan XIA, Department of Mathematics, Huazhong Normal University, Wuhan, Hubei, People's Republic of China

In this paper we generalize a result of [Turyn] and prove that there exist Hadamard matrices of orders $4t\cdot 9^i\cdot 25^j$ for $i \geq 0$, $j = 0,1,2$, and $t \in S = \{2k+1: 0 \leq k \leq 16\} \cup \{2^i\cdot 10^j\cdot 26^k+1: i,j,k \geq 0\} \cup \{37,59,61\}$.

Suppose A_1,A_2,A_3,A_4 are ±1 matrices of order n. We write $A = (A_1,A_2,A_3,A_4)$, $B = (B_1,B_2,B_3,B_4)$, etc.; we define

$$H_0(n) = \{A: \textstyle\sum_{i=1}^4 A_iA_i' = 4nI_n,\ A_iA_j = A_jA_i,\ i \neq j\},$$
$$H_1(n) = \{A: A \in H_0(n),\ A_1A_2'+A_2A_1'+A_3A_4'+A_4A_3' = 0\},$$
$$W_0(n) = \{A: A \in H_0(n),\ A_i = A_i',\ i = 1,2,3,4\}$$
$$W_1(n) = \{A: A \in W_0(n),\ A_1A_2+A_3A_4 = 0\},$$
$$W_2(n) = \{A: A \in W_1(n),\ A_1A_3+A_2A_4 = 0,\ A_1A_4+A_2A_3 = 0\}.$$

Obviously, $H_1(n) \supset W_1(n)$. It is well known that if $A \in H_0(n)$ and if the four matrices are symmetrizable by a monomial matrix R, i.e., A_iR is symmetric for $i = 1,2,3,4$, then a Goethals-Seidel type construction applies, and a Hadamard matrix of order 4n can be constructed.

THEOREM 1. (i) If $W_2(m) \neq \emptyset$ and $W_2(n) \neq \emptyset$, then $W_2(mn) \neq \emptyset$.
(ii) If $W_1(m) \neq \emptyset$ and $W_2(n) \neq \emptyset$, then $W_1(mn) \neq \emptyset$.
(iii) If $H_1(m) \neq \emptyset$ and $W_2(n) \neq \emptyset$, then $H_1(mn) \neq \emptyset$.
(iv) If $W_1(m) \neq \emptyset$ and $W_1(n) \neq \emptyset$, then $W_0(mn) \neq \emptyset$.
(v) If $H_1(m) \neq \emptyset$ and $W_1(n) \neq \emptyset$, then $H_0(mn) \neq \emptyset$.

PROOF. Suppose $A = (A_1,A_2,A_3,A_4) \in H_1(m)$, $B = (B_1,B_2,B_3,B_4) \in W_2(n)$. Put

$$C_1 = \frac{1}{2}((A_1+A_2)\times B_1 + (A_1-A_2)\times B_2),$$
$$C_2 = \frac{1}{2}((A_1+A_2)\times B_3 + (A_1-A_2)\times B_4),$$
$$C_3 = \frac{1}{2}((A_3+A_4)\times B_1 + (A_3-A_4)\times B_2),$$
$$C_4 = \frac{1}{2}((A_3+A_4)\times B_3 + (A_3-A_4)\times B_4).$$

First we see that

$$\begin{aligned}\Sigma_{i=1}^4 C_iC_i' &= \frac{1}{2}\Big[\Big[\Sigma_{i=1}^4 A_iA_i'\Big] \times \Big[\Sigma_{i=1}^4 B_iB_i'\Big] \\ &+ (A_1A_2'+A_2A_1'+A_3A_4'+A_4A_3')\times(B_1^2-B_2^2+B_3^2-B_4^2) \\ &+ 2(A_1A_1'-A_2A_2'+A_3A_3'-A_4A_4')\times(B_1B_2+B_3B_4)\Big] \\ &= 4mnI_{mn}.\end{aligned}$$

Next we have

$$C_1C_2'+C_2C_1'+C_3C_4'+C_4C_3' = 2\Big[\Sigma_{i=1}^4 A_iA_i\Big]\times(B_1B_3+B_2B_4) = 0.$$

So $C = (C_1,C_2,C_3,C_4) \in H_1(mn)$ and (iii) is proved. The proofs of other parts of Theorem 1 follow similarly.

It is shown in [1] that for any integer $i \geq 0$, $W_2(9^i) \neq \emptyset$. Theorem 2 is a generalization of this result.

THEOREM 2. For all $n = 10^i \cdot 9^j \cdot 26^k \cdot 50^r \cdot 82^s$, where $i,j,k,r,s \geq 0$, $H_1(v) \neq \emptyset$ where $v = n$ or $n/2$ according as $n \equiv 1 \pmod 2$ or not. The conclusion of Theorem 2 follows from [2] and Theorem 1.

THEOREM 3. There exist Hadamard matrices of order $4t\cdot 9^i \cdot 25^j$ for $i \geq 0$, $j = 0,1,2$ and $t \in S$.

PROOF. Let $g = x+2$ and $e_i = \{g^i, g^{i+12}\} \bmod (5,\ x^2+x+1)$, $i = 0,1,\ldots,11$. Put

$$E_1 = e_0 \cup e_1 \cup e_3 \cup e_8 \cup e_9,$$
$$E_2 = e_2 \cup e_3 \cup e_6 \cup e_7 \cup e_9,$$
$$E_3 = e_2 \cup e_5 \cup e_8 \cup e_9 \cup e_{10},$$
$$E_4 = \{0\} \cup e_0 \cup e_1 \cup e_5 \cup e_6 \cup e_7 \cup e_9 \cup e_{10}.$$

Let $h_0 = 0$, $h_{i+1} = g^i$, $i = 0,1,\ldots,23$. We define

$$A_i = \left[a_{jk}^{(i)}\right], \quad a_{jk}^{(i)} = \begin{cases} 1, & \text{if } h_k - h_j \in E_i, \\ -1, & \text{otherwise,} \end{cases} \qquad i = 1,2,3,4.$$

Thus we have A_i as the ±1 incidence matrix of E_i $(i = 1,2,3,4)$ and A_1, A_2, A_3, A_4 are symmetric, commutative, and they satisfy

$$A_1^2 + A_2^2 + A_3^2 + A_4^2 = 100 I_{25}, \quad A_1A_2 + A_3A_4 = 0.$$

So $A = (A_1, A_2, A_3, A_4) \in W_1(25)$. From Theorem 1 we know that $W_1(9^i \cdot 25) \neq \emptyset$ and $W_0(9^i \cdot 25^2) \neq \emptyset$ for any $i \geq 0$. By the same methods as Theorem 12 of [2] our proof is completed.

REFERENCES

[1] R.J. Turyn, A special class of Williamson matrices and difference sets, J. Combinatorial Theory 36A(1984), 111-115.

[2] Ming-yuan Xia, On supplementary difference sets and Hadamard matrices, Acta Mathematica Scientia 4(1984), 81-92.

Finite Vector Spaces and Block Designs

Benfu YANG, Department of Mathematics, Teacher-Training College of Chengdu, Chengdu, China

Wan-di WEI, Department of Mathematics, Sichuan University, Chengdu, China

1. INTRODUCTION

Let q be a prime power, F_q the finite field with q elements, $GL_n(F_q)$ the general linear group of degree n over F_q, and $V_n(F_q)$ the n-dimensional vector space over F_q. $GL_n(F_q)$ will be viewed as a transformation group of $V_n(F_q)$. We will first prove a theorem concerning the transitivity of $GL_n(F_q)$ on a set of some subspaces of $V_n(F_q)$, and then use it to construct a number of PBIB designs and a BIB design, the parameters of which are also computed.

2. A PROPERTY OF $GL_n(F_q)$

Let P and Q be two subspaces of $V_n(F_q)$. The $P \cup Q$ denotes the subspace which P and Q span, and is called the join of P and Q.

It is well known that $GL_n(F_q)$ acts transitively on the set of subspaces of the same dimension in $V_n(F_q)$, as well as transitively on the set of subspace pairs (X,Y) where X and Y are m-dimensional such that X ∪ Y are (m+i)-dimensional subspaces for a fixed i, $0 < i \leq \min\{m,n-m\}$ (see, for example, Chapters 1 and 6 of [2]). We have proved in [3] that $GL_n(F_q)$ acts transitively on the set of subspace pairs (X,Y) where X are m-dimensional, Y m′-dimensional, and X and Y span (m+i)-dimensional subspaces for a fixed i,

$$\max\{0,m'-m\} \leq i \leq \min\{m',n-m\}.$$

Let P be an m×n matrix with rank m over F_q. We will use the same symbol P to denote the subspace the rows of P span. Let P_0 be a given m-dimensional subspace of $V_n(F_q)$, G the subgroup consisting of the elements of $GL_n(F_q)$ which fix P_0, and W the set of 1-dimensional subspaces of $V_n(F_q)$ that are not included in P_0. Then we can prove the following:

THEOREM 1: (i) G acts transitively on W;

(ii) G acts transitively on the set of subspace pairs (x,y) where X,Y ∈ W such that X ∪ Y each intersect P_0 in the 0-dimensional subspace;

(iii) G acts transitively on the set of subspace pairs (X,Y) where x,y ∈ W such that X ∪ Y each intersect P_0 in a 1-dimensional subspace.

PROOF: (i) Let $V_1,V_2 \in W$. Then $\begin{bmatrix} P_0 \\ V_1 \end{bmatrix}$ and $\begin{bmatrix} P_0 \\ V_2 \end{bmatrix}$ are both (m+1)×n matrices with rank m+1 over F_q. Hence, there exists $T \in GL_n(F_q)$ such that $\begin{bmatrix} P_0 \\ V_1 \end{bmatrix} T = \begin{bmatrix} P_0 \\ V_2 \end{bmatrix}$, i.e.

$$P_0T = P_0, \quad V_1T = V_2$$

(see Theorem 1 in chapter 1 of [2]). Thus T ∈ G. Therefore, G acts transitively on the set W.

(ii) Let $V_1,V_2,V_3,V_4, \in W$, and

$$(V_1 \cup V_2) \cap P_0 = (V_3 \cup V_4) \cap P_0 = \{0\}.$$

Then

$$\begin{bmatrix} P_0 \\ V_1 \\ V_2 \end{bmatrix} \text{ and } \begin{bmatrix} P_0 \\ V_3 \\ V_4 \end{bmatrix}$$

are both $(m+2)\times n$ matrices with rank $m+2$ over F_q. Similarly, we have the conclusion (ii).

(iii) Let $V_1, V_2, V_3, V_4 \in W$, and

$$\dim((V_1 \cup V_2) \cap P_0) = \dim((V_3 \cup V_4) \cap P_0) = 1.$$

Write

$$(V_1 \cup V_2) \cap P_0 = aV_1 + bV_2, \quad a,b \in F_q.$$

Since $V_1, V_2 \not\subset P_0$, it follows that $ab \neq 0$. Without loss of generality, we may assume $a = 1$. Then we can write

$$P_0 = \begin{bmatrix} Q_0 \\ V_1+bV_2 \end{bmatrix},$$

where Q_0 is an $(m-1)\times n$ matrix with rank $m-1$. Since $V_1 \not\subset P_0$, it follows that

$$\begin{bmatrix} Q_0 \\ V_1+bV_2 \\ V_1 \end{bmatrix}$$

is an $(m+1)\times n$ matrix with rank $m+1$ over F_q. Similarly, without loss of generality, we may assume $(V_3 \cup V_4) \cap P_0 = V_3 + dV_4$, where $d \neq 0$. Then we can write $P = \begin{bmatrix} R_0 \\ V_3+dV_4 \end{bmatrix}$, where R_0 is an $(m-1)\times n$ matrix with rank $m-1$. Clearly,

$$\begin{bmatrix} R_0 \\ V_3+dV_4 \\ V_3 \end{bmatrix}$$

is an $(m+1)\times n$ matrix with rank $m+1$ over F_q. Therefore, there exists $T \in GL_n(F_q)$ such that

$$\begin{bmatrix} Q_0 \\ V_1+bV_2 \\ V_1 \end{bmatrix} T = \begin{bmatrix} R_0 \\ V_3+dV_4 \\ V_3 \end{bmatrix}$$

so

$$Q_0T = R_0, \tag{1}$$
$$(V_1+bV_2)T = V_3 + dV_4, \tag{2}$$
$$V_1T = V_3. \tag{3}$$

From (1) and (2) we have $P_0T = P_0$, i.e. $T \in G$; and from (2) and (3) we have $V_1T = V_3$ and $V_2T = V_4$. Hence the conclusion of (iii). This completes the proof.

3. ASSOCIATION SCHEMES

THEOREM 2. Let $0 < m < n$. Taking W as the set of treatments, and defining two treatments to be the first (resp. second) associates of each other if their join intersects P_0 in the 0-dimensional (resp. a 1-dimensional) subspace, we obtain an association scheme with two associate classes and with the parameters:

$$v = \frac{q^m(q^{n-m}-1)}{q-1}, \qquad n_1 = \frac{q^{m+1}(q^{n-m-1}-1)}{q-1},$$
$$n_2 = q^m-1, \quad p_{11}^1 = \frac{q^m(q^{n-m}-2q+1)}{q-1}, \quad p_{11}^2 = n_1. \tag{4}$$

PROOF: By the transitivity in Theorem 1, we certainly obtain an association scheme with two associate classes. We now compute its parameters.

The parameter v is the number of 1-dimensional subspaces that are not included in P_0, so

$$v = \frac{q^n-1}{q-1} - \frac{q^m-1}{q-1} = \frac{q^m(q^{n-m}-1)}{q-1}.$$

Let $V_1 \in W$. Clearly, V_2 and V_1 are the first associates of each other if and only if $V_2 \not\subseteq V_1 \cup P_0$. Noting that $\dim(V_1 \cup P_0) = m+1$, we have

$$n_1 = \frac{q^n-1}{q-1} - \frac{q^{m+1}-1}{q-1} = \frac{q^{m+1}(q^{n-m-1}-1)}{q-1}.$$

and by $n_1+n_2 = v-1$, we have

$$n_2 = v-1-n_1 = q^m-1.$$

Let V_1 and V_2 be two treatments which are the first associates of each other. Then $\dim(V_1 \cup V_2 \cup P_0) = m+2$. Let $V \in W$. If $V \not\subseteq (V_1 \cup V_2 \cup P_0)$, then V and V_i are the first associates of each other $(i = 1,2)$. The number of such V is

$$\frac{q^n-1}{q-1} - \frac{q^{m+2}-1}{q-1} = \frac{q^n-q^{m+2}}{q-1}.$$

Suppose $V \subseteq (V_1 \cup V_2 \cup P_0)$. Write $V = aV_1+bV_2+V_0$, where $a,b \in F_q$, and $V_0 \subseteq P_0$. Since $V_1 \not\subseteq (V_2 \cup P_0)$ and $V_2 \not\subseteq (V_1 \cup P_0)$, it follows that V and V_1 are the first associates of each other if and only if $b \neq 0$, and that V and V_2 are the first associates of each other if and only if $a \neq 0$. Thus, the number of treatments in $V_1 \cup V_2 \cup P_0$ which are the first associates of V_i $(i = 1,2)$ is

$$\frac{(q-1)^2 q^m}{q-1} = (q-1)q^m.$$

Therefore,

$$p_{11}^1 = \frac{q^n-q^{m+2}}{q-1} + (q-1)q^m = \frac{q^m(q^{n-m}-2q-1)}{q-1} .$$

By $p_{21}^1+p_{11}^1 = n_1-1$ and $n_1p_{21}^1 = n_2p_{11}^2$, we have

$$p_{21}^1 = n_1-1-p_{11}^1 = \frac{q^n-q^{m+1}}{q-1} - 1 = \frac{q^n-2q^{m+1}+q^n}{q-1}$$

$$= q^m-1 = n_2 ,$$

$$p_{11}^2 = \frac{n_1p_{21}^1}{n_2} = n_1 .$$

This completes the proof.

4. PBIB DESIGNS

We can construct PBIB designs based on the association scheme in Theorem 2.

THEOREM 3: Let $m < \ell < n$. Take the association scheme in Theorem 2, and take the ℓ-dimensional subspaces of $V_n(F_q)$ that include P_0 as blocks. Define a treatment to be arranged in a block if the latter includes the former both as subspaces. Then we obtain a PBIB design with two associate classes and with the parameters given in (4) and in the following.

$$b = \frac{\prod_{i=\ell-m+1}^{n-m}(q^i-1)}{\prod_{i=1}^{n-\ell}(q^i-1)} ,$$

$$k = \frac{q^m(q^{\ell-m}-1)}{q-1}, \qquad r = \frac{\prod_{i=\ell-m}^{n-m-1}(q^i-1)}{\prod_{i=1}^{n-\ell}(q^i-1)},$$

$$\lambda_1 = \frac{\prod_{i=\ell-m-1}^{n-m-2}(q^i-1)}{\prod_{i=1}^{n-\ell}(q^i-1)}, \qquad \lambda_2 = \frac{\prod_{i=\ell-m}^{n-m-1}(q^i-1)}{\prod_{i=1}^{n-\ell}(q^i-1)}.$$

PROOF: By the transitivity proved in Theorem 1, we certainly obtain a PBIB design with two associate classes. Its parameters are computed as follows.

The parameter b is the number of ℓ-dimensional subspaces of $V_n(F_q)$ that include the m-dimensional subspace P_0, so

$$b = N^T(m,\ell,n) = N(n-\ell,n-m,n) = N(n-\ell,n-m)$$
$$= \frac{\prod_{i=\ell-m+1}^{n-m}(q^i-1)}{\prod_{i=1}^{n-\ell}(q^i-1)}.$$

The parameter k is the number of 1-dimensional subspaces that are not included in P_0 and are included in a given ℓ-dimensional subspace including P_0, so

$$k = \frac{q^\ell-1}{q-1} - \frac{q^m-1}{q-1} = \frac{q^m(q^{\ell-m}-1)}{q-1}.$$

From these, the parameter r is obtained:

$$r = \frac{bk}{v} = \frac{\prod_{i=\ell-m}^{n-m-1}(q^i-1)}{\prod_{i=1}^{n-\ell}(q^i-1)}.$$

The parameter λ_1 is the number of ℓ-dimensional subspaces that include P_0 and two given 1-dimensional subspaces which are the first associates of each other, i.e., the number of ℓ-dimensional subspaces that include a given (m+2)-dimensional subspace, so

$$\lambda_1 = \frac{\prod_{i=\ell-m-1}^{n-m-2} (q^i-1)}{\prod_{i=1}^{n-\ell} (q^i-1)} .$$

The parameter λ_2 is the number of ℓ-dimensional subspaces that include a given (m+1)-dimensional subspace, so

$$\lambda_2 = \frac{\prod_{i=\ell-m}^{n-m-1} (q^i-1)}{\prod_{i=1}^{n-\ell} (q^i-1)} .$$

It can also be derived from $r(k-1) = \lambda_1 n_1 + \lambda_2 n_2$. This completes the proof.

THEOREM 4: Take the association scheme given in Theorem 2. Let $0 < \ell \leq n-m$. Taking as blocks the ℓ-dimensional subspaces of $V_n(F_q)$ that each with P_0 span $(m+\ell)$-dimensional subspaces, and defining a treatment to be arranged in a block if the latter includes the former both as subspaces, we obtain a PBIB design with two associate classes and with the parameters given in (4) and in the following

$$b = \frac{\prod_{i=n-m-\ell+1}^{n-m} (q^i-1)}{\prod_{i=1}^{\ell} (q^i-1)} q^{\ell m},$$

$$k = \frac{q^\ell-1}{q-1}, \qquad r = \frac{bk}{v},$$

$$\lambda_2 = 0, \qquad \lambda_1 = \frac{r(k-1)}{n_1} .$$

PROOF: By the transitivity of G in Theorem 1, we certainly obtain a PBIB design with two associate classes. The parameter b has been computed in (3), and the other parameters can be computed immediately. This proves the theorem.

THEOREM 5: Take the association scheme in Theorem 2. Let $2 \leq \ell \leq m$. Taking as blocks the ℓ-dimensional subspaces that each with P_0 span (m+1)-dimensional subspaces, and defining a treatment to be arranged in a block if the latter includes the former both as subspaces, we obtain a PBIB design with two associate classes and with the parameters given in (4) and in the following

$$b = \frac{(q^{n-m}-1)\prod_{i=m-\ell+2}^{n-m}(q^i-1)}{(q-1)\prod_{i=1}^{\ell-1}(q^i-1)}\, q^{m-\ell+1},$$

$$k = q^{\lambda-1}, \qquad r = \frac{bk}{v},$$

$$\lambda_1 = 0, \qquad \lambda_2 = \frac{r(k-1)}{n_2}.$$

PROOF: By the transitivity in Theorem 1, we certainly obtain a PBIB design with two associate classes. The parameter b has been computed in [3]. The parameter k is the number of 1-dimensional subspaces that are not included in P_0 and are included in a given ℓ-dimensional subspace which with P_0 spans an (m+1)-dimensional subspace, so

$$k = N(1,\ell) - N(1,\ell-1) = q^{\ell}-1.$$

and the other parameters are computed immediately. This proves the Theorem.

THEOREM 6. Let $m = n-2$. Take the association scheme given in Theorem 2. Let $2 \leq \ell \leq m+1$. Taking as blocks the ℓ-dimensional subspaces that each with P_0 span the total space $V_n(F_q)$, and defining a treatment to be arranged in a block if the latter includes the former both as subspaces, we obtain a PBIB design with two associate classes and with the parameters given in (4) and in the following

$$b = \frac{\prod_{i=n-\ell+1}^{n-2}(q^i-1)}{\prod_{i=1}^{\ell-2}(q^i-1)}\, q^{2(n-\ell)},$$

$$k = q^{\ell-2}(q+1), \qquad r = \frac{bk}{v},$$

$$\lambda_1 = \frac{\prod_{i=\ell-1}^{n-2}(q^i-1)}{\prod_{i=1}^{n-\ell}(q^i-1)}, \qquad \lambda_2 = \frac{r(k-1)-\lambda_1 n_1}{n_2}.$$

PROOF: By the transitivity in Theorem 1, we certainly obtain a PBIB design with the two associate classes. The parameter b has been computed in [3]. The parameter λ_1 can be computed as follows. Let V_1 and V_2 be two treatments that are the first associates of each other. Then $V_1 \cup V_2 \cup P_0 = V_n(F_q)$, and P_0 and any ℓ-dimensional subspace that includes V_1 and V_2 span $V_n(F_q)$. Thus

$$\lambda_1 = N^T(2,\ell,n) = N(n-\ell,n-2) = \frac{\prod_{i=\ell-1}^{n-2}(q^i-1)}{\prod_{i=1}^{n-\ell}(q^i-1)}.$$

The other parameters are computed immediately. This proves the Theorem.

THEOREM 7: Take the association scheme in Theorem 2. Taking the treatments also as blocks, and defining a treatment to be arranged in a block if the subspace spanned by them (as 1-dimensional subspaces) intersects P_0 in the 0-dimensional subspace, we obtain a PBIB design with two associate classes and with the parameters given in (4) and in the following

$$b = v, \qquad k = n_1, \qquad r = k,$$
$$\lambda_1 = p_{11}^1, \qquad \lambda_2 = p_{11}^2.$$

The theorem is obviously true.

If we put $m = n-1$ in Theorem 2, then $n_1 = 0$, and a BIB design can be constructed as follows.

THEOREM 8: Let P_0 be a given $(n-1)$-dimensional subspace, and $1 < \ell < n$. The 1-dimensional subspaces of $V_n(F_q)$ that are not included in P_0 are taken as treatments. The ℓ-dimensional subspaces that are not included in P_0 are taken as blocks. And a treatment is defined to be arranged in a block if the latter includes the former both as subspaces. Then we obtain a BIB design with the parameters

$$b = \frac{\prod_{i=n-\ell+1}^{n-1}(q^i-1)}{\prod_{i=1}^{\ell-1}(q^i-1)} q^{n-\ell},$$

$$v = q^{n-1}, \qquad k = q^{\ell-1},$$

$$r = \frac{\prod_{i=\ell}^{n-1}(q^i-1)}{\prod_{i=1}^{n-\ell}(q^i-1)}, \qquad \lambda = \frac{\prod_{i=\ell-1}^{n-2}(q^i-1)}{\prod_{i=1}^{n-\ell}(q^i-1)}.$$

PROOF: By the transitivity in Theorem 1, we certainly obtain a BIB design. We now compute the parameters.

The parameter b is the number of ℓ-dimensional subspaces minus the number of ℓ-dimensional subspaces that are included in P_0, i.e.,

$$b = N(\ell,n) - N(\ell,n-1) = \frac{\prod_{i=n-\ell+1}^{n-1}(q^i-1)}{\prod_{i=1}^{\ell-1}(q^i-1)}.$$

Similarly,

$$v = N(1,n) - N(1,n-1) = q^{n-1}.$$

Since an ℓ-dimensional subspace which is a block intersects P_0 in an $(\ell-1)$-dimensional subspace, it follows that

$$k = N(1,\ell) - N(1,\ell-1) = q^{\ell-1}.$$

The parameter r is the number of ℓ-dimensional subspaces that include a given 1-dimensional subspace. By the conjugation among the subspaces (see [4]), we have

$$r = N^T(1,\ell,n) = N(n-\ell,n-1) = \frac{\prod_{i=\ell}^{n-1}(q^i-1)}{\prod_{i=1}^{n-\ell}(q^i-1)}.$$

It can also be derived from the relation rv = bk. The parameter λ is the number of ℓ-dimensional subspaces that include a given 2-dimensional subspace:

$$\lambda = N^T(2,\ell,n) = N(n-\ell,n-2) = \frac{\prod_{i=\ell-1}^{n-2}(q^i-1)}{\prod_{i=1}^{\ell-1}(q^i-1)} .$$

It can also be derived from the relation $\lambda(v-1) = r(k-1)$. This proves the theorem.

5. CONSTRUCTING PBIB DESIGNS BY USING THE CONGRUENT CLASSES OF ALTERNATING MATRICES, SYMMETRIC MATRICES AND HERMITIAN MATRICES

In this section, we will use the congruent classes of nonsingular alternating matrices, the congruent classes of nonsingular symmetric matrices and the congruent classes of nonsingular Hermitian matrices to construct a number of PBIB designs.

First of all, we discuss the case of alternating matrices.

Let $n = 2\nu$, $\nu > m$. Take the association scheme in Theorem 2. The congruent classes of nonsingular alternating $2\nu \times 2\nu$ matrices over F_q that have P_0 as their totally isotropic subspace are taken as blocks. $\mathcal{B}$ denotes the set of blocks. A treatment V is said to be arranged in a block if the matrices in the congruent class that has been taken as the block have $\begin{bmatrix} P_0 \\ V \end{bmatrix}$ as their totally isotropic (m+1)-dimensional subspace. We will now prove that we obtain a PBIB design with two associate classes.

We first prove that the number k of treatments in a block is a constant not depending on the choice of the block. Let K be a matrix of a congruent class $B \in \mathcal{B}$. Then k is the number of 1-dimensional subspaces in P_0^* (the conjugate subspace of P_0 with respect to K) but not in P_0. Since $P_0 \subset P_0^*$ and $\dim P^* = 2\nu-m$, it follows that

$$k = \frac{q^{2\nu-m}-1}{q-1} - \frac{q^m-1}{q-1} = \frac{q^m(q^{2\nu-2m}-1)}{q-1}$$

a constant not depending on the choice of the block.

On the other hand, by the transitivity in Theorem 1, the parameters r, λ_1 and λ_2 are all constants. Thus we certainly obtain a PBIB design with two associate classes, and we only need to compute the values of b, r, λ_1 and λ_2.

Take

$$P_0 = (I^{(m)}, 0^{(m,2\nu-m)}).$$

Then $\mathcal{B}$ consists of the congruent classes of nonsingular alternating matrices of the form

$$\begin{matrix} \begin{bmatrix} 0 & K_{12} \\ -K_{12}' & K_{22} \end{bmatrix} & \begin{matrix} m \\ 2\nu-m \end{matrix} \\ \begin{matrix} m & \quad 2\nu-m \end{matrix} & \end{matrix} \tag{5}$$

It is known (see Lemma 4 in Chapter 6 of [2]) that the number of nonsingular alternating matrices of the form (5) is

$$\prod_{i=2\nu-2m+1}^{2\nu-m}(q^i-1)\prod_{i=1}^{\nu-m}(q^{2i-1}-1)q^{\nu(\nu-1)} \tag{6}$$

so

$$b = \prod_{i=2\nu-2m+1}^{2\nu-m}(q^i-1)\prod_{i=2}^{\nu-m}(q^{2i-1}-1)q^{\nu(\nu-1)}.$$

By the relation bk = rv, we have

$$r = \frac{bk}{v} = \frac{\prod_{i=2\nu-2m+1}^{2\nu-m}(q^i-1)\prod_{i=2}^{\nu-m}(q^{2i-1}-1)q^{\nu(\nu-1)}q^m(q^{2\nu-2m}-1)(q-1)}{(q-1)q^m(q^{2\nu-m}-1)}$$

$$= \prod_{i=2\nu-2m}^{2\nu-m-1}(q^i-1)\prod_{i=2}^{\nu-m}(q^{2i-1}-1)q^{\nu(\nu-1)}.$$

In order to compute the value of λ_2, take $V_1 = e_{m+1} =$ (0,...,0,1,0,...,0) (m leading zeroes) and $V_2 = e_1+e_{m+1}$. Clearly, V_1 and V_2 are the second associates of each other, and the nonsingular alternating matrices that have $\begin{bmatrix} P_0 \\ V_1 \end{bmatrix}$ as their totally isotropic subspace will have $\begin{bmatrix} P_0 \\ V_2 \end{bmatrix}$ as their totally isotropic subspace. Replacing m by m+1 in (6), we find the number of nonsingular alternating matrices that have $\begin{bmatrix} P_0 \\ V_1 \end{bmatrix}$ as its totally isotropic subspace to be

$$\prod_{i=2\nu-2m-1}^{2\nu-m-1} (q^i-1) \prod_{i=1}^{\nu-m-1} (q^{2i-1}-1) q^{\nu(\nu-1)}$$

so

$$\lambda_2 = \prod_{i=2\nu-2m-1}^{2\nu-m-1} (q^i-1) \prod_{i=2}^{\nu-m-1} (q^{2i-1}-1) q^{\nu(\nu-1)}.$$

The parameter λ_1 follows from $r(k-1) = \lambda_1 n_1 + \lambda_2 n_2$:

$$\lambda_1 = \frac{1}{n_1}[r(k-1)-\lambda_2 n_2]$$

$$= \frac{q-1}{q^{m+1}(q^{2\nu-m-1}-1)} \left[\prod_{i=2\nu-2m}^{2\nu-m-1} (q^i-1) \prod_{i=2}^{\nu-m} (q^{2i-1}-1) q^{\nu(\nu-1)} \right.$$

$$\left. \cdot \frac{q^{2\nu-m}-q^m-q+1}{q-1} - \prod_{i=2\nu-2m-1}^{2\nu-m-1} (q^i-1) \prod_{i=2}^{\nu-m-1} (q^{2i-1}-1) q^{\nu(\nu-1)} (q^m-1) \right]$$

$$= \prod_{i=2\nu-2m-1}^{2\nu-m-2} (q^i-1) \prod_{i=2}^{\nu-m} (q^{2i-1}-1) q^{\nu(\nu-1)}.$$

Therefore, we have proved

THEOREM 9: Let $n = 2\nu$, $\nu > m$, and P_0 be a given m-dimensional subspace of $V_{2\nu}(F_q)$. Take the association scheme in Theorem 2. Taking as blocks the congruent classes of nonsingular alternating $2\nu \times 2\nu$ matrices over F_q that have P_0 as their totally isotropic subspace, and defining a treatment V to be arranged in a block B if the matrices in the congruent class B have $\begin{bmatrix} P_0 \\ V \end{bmatrix}$ as their totally isotropic (m+1)-dimensional subspace, we obtain a PBIB design with two associate classes and with the parameters given in (4) and in the following

$$b = \prod_{i=2\nu-2m+1}^{2\nu-m} (q^i-1) \prod_{i=2}^{\nu-m} (q^{2i-1}-1) q^{\nu(\nu-1)}$$

$$k = \frac{q^{2\nu-2m}-1}{q-1}\, q^m,$$

$$r = \prod_{i=2\nu-2m}^{2\nu-m-1} (q^i-1) \prod_{i=2}^{\nu-m} (q^{2i-1}-1)\, q^{\nu(\nu-1)},$$

$$\lambda_1 = \prod_{i=2\nu-2m-1}^{2\nu-m-2} (q^i-1) \prod_{i=2}^{\nu-m} (q^{2i-1}-1)\, q^{\nu(\nu-1)},$$

$$\lambda_2 = \prod_{i=2\nu-2m-1}^{2\nu-m-1} (q^i-1) \prod_{i=2}^{\nu-m-1} (q^{2i-1}-1)\, q^{\nu(\nu-1)}.$$

Next, we discuss the case of symmetric matrices.

THEOREM 10: Let $n = 2\nu+\delta$ ($\delta = 0,1$ or 2), $0 < m < \nu$, and P_0 be a given m-dimensional subspace. Take the association scheme in Theorem 2. Taking as blocks the congruent classes of nonsingular symmetric $n\times n$ matrices over F_q that have P_0 as their totally isotropic subspace, and defining a treatment V to be arranged in a block B if the matrices in the congruent class B have $\begin{bmatrix} P_0 \\ V \end{bmatrix}$ as their totally isotropic (m+1)-dimensional subspace, we obtain a PBIB design with two associate classes and with the parameters given in (4) and in the following

$$b = \begin{cases} \frac{1}{2}(q^{\nu-m}+1) \prod_{i=2\nu-2m+1}^{2\nu-m} (q^i-1) \prod_{i=1}^{\nu-m-1} (q^{2i+1}-1)\, q^{\nu^2}, & \text{if } \delta = 0, \\ \prod_{i=2\nu-2m+2}^{2\nu+1-m} (q^i-1) \prod_{i=1}^{\nu-m} (q^{2i+1}-1)\, q^{\nu(\nu-1)}, & \text{if } \delta = 1, \\ \frac{1}{2}(q^{\nu-m+1}-1) \prod_{i=2\nu-2m+3}^{2\nu+2-m} (q^i-1) \prod_{i=1}^{\nu-m} (q^{2i+1}-1)\, q^{(\nu+1)^2}, & \text{if } \delta = 2 \end{cases} \tag{7}$$

$$k = \frac{(q^{\nu-m}-1)(q^{\nu-m+\delta-1}+1)}{q-1}\, q^m, \qquad r = \frac{bk}{v},$$

$$\lambda_2 = \begin{cases} \frac{1}{2}(q^{\nu-m-1}+1)\prod_{i=2\nu-2m-1}^{2\nu-m-1}(q^i-1)\prod_{i=1}^{\nu-m-2}(q^{2i+1}-1)q^{\nu^2}, & \text{if } \delta = 0, \\ \prod_{i=2\nu-2m}^{2\nu-m}(q^i-1)\prod_{i=1}^{\nu-m-1}(q^{2i+1}-1)q^{\nu(\nu+1)}, & \text{if } \delta = 1, \\ \frac{1}{2}(q^{\nu-m}-1)\prod_{i=2\nu-2m+1}^{2\nu-m+1}(q^i-1)\prod_{i=1}^{\nu-m-1}(q^{2i+1}-1)q^{(\nu+1)^2}, & \text{if } \delta = 2. \end{cases} \quad (8)$$

$$\lambda_1 = \frac{1}{n_1}[r(k-1) - \lambda_2 n_2].$$

PROOF: In the same way as proving Theorem 9, we know that the obtained is certainly a PBIB design with two associate classes. Its parameters can be computed as follows.

We know (see [2]) that in the $(2\nu+\delta)$-dimensional orthogonal geometry, the conjugate subspace of an m-dimensional totally isotropic subspace is of $(2\nu-m+\delta, 2(\nu-m)+\delta, \nu-m, \Delta; 2\nu+\delta, \Delta)$-type. Then by a similar reason to that in Theorem 9, the parameter k is

$$k = N(1,0,0;\ 2\nu-m+\delta,\ 2(\nu-m)+\delta,\ \nu-m,\ \Delta;\ 2\nu+\delta,\ \Delta) - \frac{q^m-1}{q-1}$$

$$= \frac{(q^{\nu-m}-1)(q^{\nu-m+\delta-1}+1)}{q-1}\, q^m.$$

In order to compute the values of the parameters b, r, λ_1 and λ_2, take

$$P_0 = (I^{(m)}, 0^{(m,n-m)}).$$

Then it is easy to see that the blocks are the congruent classes of nonsingular symmetric matrices of the form

$$\begin{bmatrix} 0 & S_{12} \\ S'_{12} & S_{22} \end{bmatrix} \begin{matrix} m \\ n-m \end{matrix} \qquad (9)$$
$$\begin{matrix} m \quad & n-m \end{matrix}$$

From the number of nonsingular symmetric matrices of the form (9), which was given in Lemma 8 of Chapter 6 of [2], (7) follows. By replacing m by m+1 in (7), (8) follows. This completes the proof.

Finally, we discuss the case of Hermitian matrices.

Let P_0 be a given m-dimensional subspace of $V_n(F_{q^2})$, where $m < \left[\frac{n}{2}\right]$. Let U be the set of 1-dimensional subspaces of $V_n(F_{q^2})$ that are not included in P_0. Then we can prove

THEOREM 11: Take U as the set of treatments, and define two treatments to be the first (resp. second) associates of each other if the two treatments as subspaces span a subspace which intersects P_0 in the 0-dimensional (resp. a 1-dimensional) subspace. Taken as blocks the congruent classes of nonsingular Hermitian n×n matrices that have P_0 as their totally isotropic subspace, and define a treatment V to be arranged in a block B if the matrices in the congruent class B have $\begin{bmatrix} P_0 \\ V \end{bmatrix}$ as their totally isotropic (m+1)-dimensional subspace. Then we obtain a PBIB design with two associates and with the parameters:

$$v = \frac{q^{2m}(q^{2n-2m}-1)}{q^2-1}, \qquad n_1 = \frac{q^{2(m+1)}(q^{2n-2m-2}-1)}{q^2-1},$$

$$n_2 = q^{2m}-1, \qquad p_{11}^1 = \frac{q^{2m}(q^{2n-2m}-2q^2+1)}{q^2-1}, \qquad p_{11}^2 = n_1,$$

$$b = \frac{1}{q^2-1} \prod_{i=n-2m+1}^{n-m} (q^{2i}-1) \prod_{i=1}^{n-2m} (q^i+(-1)^i) q^{\frac{n(n-1)}{2}},$$

$$k = \frac{1}{q^2-1}(q^{n-2m}-(-1)^{n-2m})(q^{n-2m-1}-(-1)^{n-2m-1})q^{2m},$$

$$r = \frac{1}{q^2-1} \prod_{i=n-2m-1}^{n-m-1} (q^{2i}-1) \prod_{i=1}^{n-2m} (q^i+(-1)^i) q^{\frac{n(n-1)}{2}},$$

$$\lambda_2 = \frac{1}{q^2-1} \prod_{i=n-2m-1}^{n-m-1} (q^{2i}-1) \prod_{i=1}^{n-2m-2} (q^i+(-1)^i) q^{\frac{n(n-1)}{2}},$$

$$\lambda_1 = \frac{1}{n_1}[r(k-1) - \lambda_2 n_2].$$

PROOF: Let B be a congruent class of nonsingular Hermitian n×n matrices, and $H \in B$. Then the number of treatments in the block B is the number of 1-dimensional isotropic subspaces which are included in P_0^* (the conjugation of P_0 with respect to H) but are not included in P_0. We know (see [4]) that P^* is an (n-m,n-2m)-type subspace. Noting that $P_0 \subseteq P_0^*$, and that the 1-dimensional subspaces of P_0 are isotropic, we have

$$k = N(1,0;\ n-m,\ n-2m;\ n) - \frac{q^{2m}-1}{q^2-1}$$
$$= \frac{(q^{n-2m}-(-1)^{n-2m})(q^{n-2m-1}-(-1)^{n-2m-1})}{q^2-1}\, q^{2m}$$

which is a constant not depending on the choice of B.

By the transitivity in Theorem 1, the parameters r, λ_1 and λ_2 are also constants. So we certainly obtain a PBIB design with two associate classes. The parameters b, r, λ_1 and λ_2 can be computed as follows.

Take

$$P_0 = (I^{(m)}, 0^{(m,n-m)}).$$

Then it is easy to see that the blocks are the congruent classes of nonsingular Hermitian matrices of the form

$$\begin{array}{cc} \begin{bmatrix} 0 & H_{12} \\ \bar{H}'_{12} & H_{22} \end{bmatrix} & \begin{array}{l} m \\ n-m \end{array} \\ \begin{array}{cc} m & \quad n-m \end{array} & \end{array} \tag{10}$$

From the number of nonsingular Hermitian matrices of the form (10), which was given in Lemma 6 of Chapter 6 of [2], the value of b follows:

$$b = \frac{1}{q^2-1} \prod_{i=n-2m+1}^{n-m} (q^{2i}-1) \prod_{i=1}^{n-2m} (q^i+(-1)^i)\, q^{\frac{n(n-1)}{2}}. \tag{11}$$

By the relation rv = bk, we have

$$r = \frac{bk}{v} = \frac{1}{q^2-1} \prod_{i=n-2m-1}^{n-m-1} (q^{2i}-1) \prod_{i=1}^{n-2m-2} (q^i+(-1)^i)\, q^{\frac{n(n-1)}{2}}.$$

Replacing m by m+1 in (11), we have

$$\lambda_2 = \frac{1}{q^2-1} \prod_{i=n-2m-1}^{n-m-1} (q^{2i}-1) \prod_{i=1}^{n-2m-2} (q^i+(-1)^i)\, q^{\frac{n(n-1)}{2}}.$$

And λ_1 follows from the relation $r(k-1) = \lambda_1 n_1 + \lambda_2 n_2$. This completes the proof.

REFERENCES

[1] Hao Shen, Using finite geometries to construct block designs, J. Shanghai Jiao Tong University, 21(1987), No. 2, 27-37.

[2] Zhexian Wan, Zongduo Dai, Xuning Feng and Benfu Yang, Some studies on finite geometries and incomplete block designs, Science Press, Beijing, 1966.

[3] Wandi Wei and Benfu Yang, Construction of PBIB designs by using subspaces of vector spaces (to appear).

[4] Benfu Yang and Wandi Wei, Finite unitary geometries and PBIB designs (I), J. Comb. Math. Comb. Comp. 6(1989), 51-61.

Existence of G-Designs with $|V(G)| = 6$

YIN Jianxing, Suzhou University, Suzhou, China

GONG Busheng, Suzhou University, Suzhou, China

1. INTRODUCTION

Let G be a simple graph without isolated vertices. A graph H is said to be G-decomposable (or admit a G-decomposition), denoted by $H \to G$, if the edges of H can be partitioned into edges of subgraphs isomorphic to G. A G-design is a G-decomposition of K_n, the complete graph with n vertices. If G is K_k, a G-design $K_n \to G$ is nothing more than a (n,k,1)-BIBD (balanced incomplete block design). G-designs can also be defined by using the concept of G-blocks, similar to the corresponding definition of a BIBD (see [5]).

Simple calculation shows that if $K_n \to G$ then

$$\left.\begin{array}{ll} \text{(i)} & n \geq k \\ \text{(ii)} & n(n-1) \equiv 0 \pmod{2e} \\ \text{(iii)} & n-1 \equiv 0 \pmod{f} \end{array}\right\} \qquad (1.1)$$

where G is a graph with k vertices and e edges and f is the greatest common divisor of the degrees of the vertices of G.

The conditions for the existence of a G-design have been studied extensively (for example see [2]). The case when G is of at most four vertices has been solved completely in [1]. When a graph with five vertices, an almost complete solution has been given in [3].

In this paper we consider the case when G is of six vertices and e edges, where $3 \leq e \leq 6$. As we see in Figure 1, there are 28 non-isomorphic graphs (from [7] Appendix I) which need to be considered. It will be shown that the necessary condition in (1.1) for the existence of G-designs for these graphs is also sufficient with five exceptions for which G-designs do not exist.

We would like to remark that for undefined graph-theoretical concepts and notations one should see [7]. In what follows we always assume that V(G) and E(G) are the vertex-set and the edge-set of G respectively. We always denote the graph G in Figure 1 by $\langle o_1, o_2, \ldots, o_6 \rangle$.

2. PRELIMINARIES

In this section, we shall define some terminology and state some fundamental results which will be useful later on. Furthermore, we shall also use GDDs (group divisible designs) to establish a recursive construction for G-designs.

DEFINITION 2.1. A t-system is a set of disjoint pairs (p_r, q_r), $r = 1,2,\ldots,t$, such that $q_r - p_r = r$ for each r and $\bigcup_{r=1}^{t}\{p_r, q_r\} = \{1,2,\ldots,2t-1,2t+\epsilon_t\}$, where $\epsilon_t = 0$ if $t \equiv 0,1 \pmod 4$ and $\epsilon_t = 1$ otherwise.

The following Lemma can be found in [9].

LEMMA 2.2. There exists a t-system for each t.

DEFINITION 2.3. Let X be a set of v points. Let $\mathcal{A}$ be a collection of some k-subsets (called blocks) of X. A balanced incomplete block design (BIBD) of index 1 is a pair $(X,\mathcal{A})$ such that every unordered pair of points is contained in a unique block in $\mathcal{A}$. We denote $(X,\mathcal{A})$ by (v,k,1)-BIBD.

FIGURE 1

Graphs with $|V(G)| = 6$ and $3 \leq |E(G)| \leq 6$

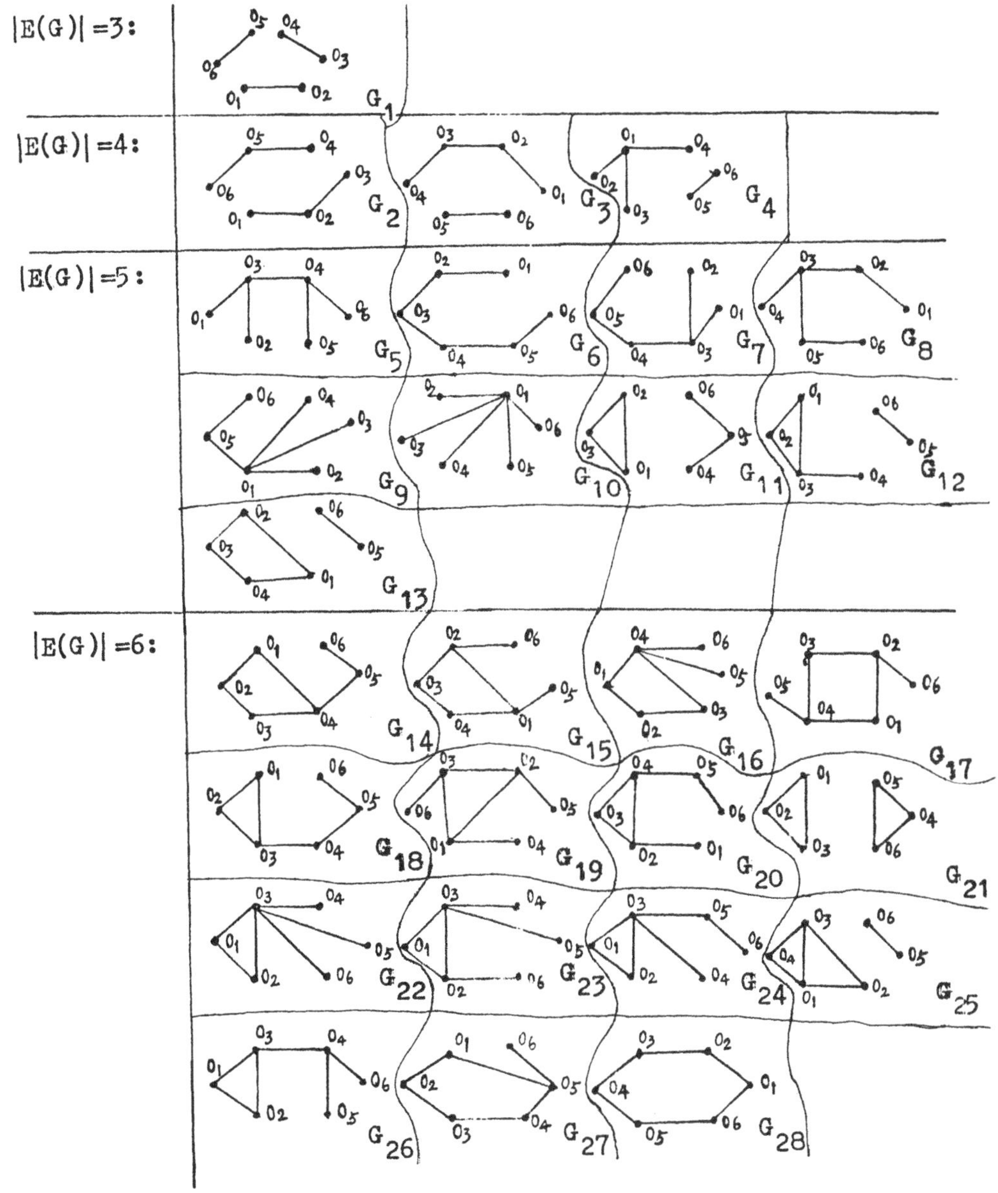

DEFINITION 2.4. Let $(X,\mathcal{A})$ be a $(v,k,1)$-BIBD. A parallel class in $(X,\mathcal{A})$ is a collection of disjoint blocks of $\mathcal{A}$, the union of which is X. $(X,\mathcal{A})$ is called resolvable if the blocks of $\mathcal{A}$ can be partitioned into parallel classes, and denoted by $(v,k,1)$-RBIBD.

DEFINITION 2.5. Let K and M be sets of positive integers. A group divisible design (GDD), denoted by $GD[K,1,M;v]$, is a triple $(X,\mathcal{G},\mathcal{A})$ satisfying:

(1) X is a v-set (of points);

(2) $\mathcal{G}$ is a collection of non-empty subsets of X (called groups) with sizes in M and which partition X;

(3) $\mathcal{A}$ is a collection of subsets of X (called blocks), each with size at least two in K;

(4) no block meets any group in more than one point;

(5) each pair of points $\{x,y\}$ not contained in a group is contained in exactly one block.

We shall write $GD[k,1,m;v]$ for $GD[\{k\},1,\{m\};v]$. The following is well known (see [6]).

LEMMA 2.6. There are $(v,3,1)$-RBIBD for all positive integers $v \equiv 3 \pmod 6$.

Hanani [5] has proved the following.

LEMMA 2.7. A necessary and sufficient condition for the existence of a $GD[3,1,m;um]$ is that $u \geq 3$, $m(u-1) \equiv 0 \pmod 2$ and $um^2(u-1) \equiv 0 \pmod 6$.

LEMMA 2.8. Suppose that there is a $GD[k,1,m;um]$. If a $K_{r,\ldots,r} \to G$ (k copies of r) and a $K_{r,\ldots,r,s_i} \to G$ (k-1 copies of r) ($s_i > 0$, $i = 1,2$) exist, then a $K_{mr,\ldots,mr,as_1+bs_2} \to G$ (u-1 copies of mr) exists, where $0 \leq a, b \leq m$ and $a+b = m$.

PROOF. The existence of a $K_{m,\ldots,m} \to K_k$ (u copies of m) is equivalent to that of a $GD[k,1,m;um]$. In a $GD[k,1,m;um]$, we can give weight s_1 to a of the points in one group and weight s_2 to the remaining b points of the group. Give weight r to the other points of the GDD. It follows that the edges of $K_{mr,\dots,mr,as_1+bs_2}$ (u-1 copies of mr) can be partitioned into edges of subgraphs isomorphic to either $K_{r,r,\dots,r,s_1}$ or $K_{r,r,\dots,r,s_1}$ or $K_{r,r,\dots,r}$ (k-1 copies of r). Then the conclusion follows from the hypothesis.

We need also the following two Lemmas for which the reader should see [2].

LEMMA 2.9. If $K_{n_i+q} \to G$ for $1 \leq i \leq h$ and $K_{n_1,\dots,n_h} \to G$, then $K_n \to G$, where $n = q + \sum_{i=1}^{h} n_i$ and $q = 0$ or 1.

LEMMA 2.10. If $K_{r_1,r_2} \to G$ and $K_{r_1,r_2'} \to G$, then $K_{ar_1,br_2+cr_2'} \to G$ for integers $a \geq 1$ and b or $c \geq 1$.

3. CONSTRUCTIONS FOR G-DESIGNS

In this section, we shall find necessary and sufficient conditions for $K_n \to G$, where G is in Figure 1. Two methods will be used: the method of differences and the method of composition. We assume that the reader is familiar with Bose's method of symmetrically repeated differences (see [4]). When we use the method of differences to construct $K_n \to G$, we will give only the "base graphs" (see [2]) of the decomposition since the rest of the graphs can be obtained by applying an automorphism of the group Γ on the vertices of the base graphs. Unless otherwise stated we always take $V(K_n) = Z_n$ or $Z_{n-1} \cup \{\infty\}$, where Z_n is the group of the residues of integers mod n. We always assume that $n \geq 6$. We shall also adapt the following notations:

$$B(G) = \{n | K_n \to G\}, \quad I_m = \{1,2,\dots,m\}.$$

THEOREM 3.1. $n \in B(G_1)$ iff $n \equiv 0$ or 1 (mod 3).

PROOF. The condition $n \equiv 0$ or 1 (mod 3) is necessary by (1.1). To prove the sufficiency, we consider four cases.

(i) $n = 6t$ and $V(K_n) = Z_{n-1} \cup \{\infty\}$; $K_n \to G_1$: the subgraphs of the decomposition are

$B_j = \langle 0,3j-2,2,3j+1,3,3j+3\rangle$ (mod n-1), $1 \le j \le t-1$;
$B_t = \langle 0,3t-1,1,3t+2,3t+1,\infty\rangle$ (mod n-1).

(ii) $n = 6t+1$ and $V(K_n) = Z_n$, $K_n \to G_1$: $\langle 0,3j-2,2,3j+1,3,3j+3\rangle$ (mod n), where $1 \le j \le t$.

(iii) $n = 6t+3$, $V(K_n) = Z_n$, $K_n \to G_1$: $\langle 0,3j-2,2,3j+1,3,3j+3\rangle$, $\langle 0,3t+1,2t+1,5t+2,t,4t+2\rangle^*$(mod n), where $1 \le j \le t$; the star * means that the repeated blocks from the base block will be taken only once.

(iv) $n = 6t+4$, $V(K_n) = Z_n$, $K_n \to G_1$: $\langle 0,3j-2,2,3j+1,3,3j+3\rangle$, $\langle 0,3t+1,1,3t+3,3t+2,6t+3\rangle^*$ (mod n), where $1 \le j \le t$.

Hence the proof is completed.

THEOREM 3.2. For $i = 2,3,4$, $n \in B(G_i)$ iff $n \equiv 0$ or 1 (mod 8).

PROOF. Again, the necessary condition follows from (1.1). For the sufficiency, we consider the following two cases.

(i) $n = 8t+1$, $K_n \to G_2$: $\langle 0,j,2t+1,4t+1+j,t+1,4t+2-j\rangle$ (mod n);
$K_n \to G_3$: $\langle 0,j,2t+1,5t+2-j,4t,7t+j\rangle$ (mod n);
$K_n \to G_4$: $\langle 0,j,2t+1-j,3t+1-j,3t+1,6t+1+j\rangle$ (mod n),
where $1 \le j \le t$.

(ii) $n = 8t$, $K_n \to G_2$: $B_j = \langle 0,j,2t-1,4t-3+j,t,4t-2-j\rangle$ and
$B_t = \langle 0,4t-1,8t-3,1,4t-2,\infty\rangle$ (mod n-1);
$K_n \to G_3$: $B_j = \langle 0,j,2t-1,5t-3-j,3t-3,6t-6+j\rangle$, and
$B_t = \langle 4t-1,0,4t-2,1,5,\infty\rangle$ (mod n-1);
$K_n \to G_4$: $B_j = \langle 0,j,2t-1-j,3t-2-j,3t-2,6t-5+j\rangle$ and
$B_t = \langle 0,4t-1,4t-2,4t-3,4,\infty\rangle$ (mod n-1), where $1 \le j \le t-1$.

The following theorem is due to C. Huang and A. Rosa [9].

THEOREM 3.3. $v \in B(G_i)$ iff $v \equiv 0,1$ (mod 5) for $i = 5,6,8$ and $v \equiv 0,1$ (mod 5) and $v > 6$ for $i = 7,9,10$.

LEMMA 3.4. $6 \notin B(G_{11})$.

PROOF. Assume that $6 \in B(G_{11})$, i.e., we can decompose K_6 into three subgraphs, B_1,B_2,B_3, which are isomorphic to G_{11}. The G_{11} has four vertices of degree 2, two vertices of degree 1. Each vertex in K_6 has degree 5, of course. Thus each vertex in K_6

must have type (2,2,1) (a vertex has type (x,y,z) if x,y,z are its degrees in the 3 subgraphs of the decomposition). Without loss of generality, we assume that $V(K_6) = Z_6$ and $B_1 = \langle 0,1,2,3,4,5\rangle$. Then the vertex 0 must appear as a vertex of degree 2 in B_2 or B_3, say, in B_2. We are forced to have $B_2 = \langle 0,3,5,1,4,2\rangle$. But the remaining edges $E(K_6)\setminus(E(B_1) \cup E(B_2))$ do not contain any triangle and actually form a chain; this is a contradiction.

LEMMA 3.5. If $n \equiv 0,1 \pmod 5$ and $n > 6$, then $n \in B(G_j)$ for $j = 11,12$.

PROOF. (1) $n = 10t+1$, $K_n \to G_{11}$: $B_1 = \langle 0,t+p_1,t+q_1,4t,8t+1,5t+\epsilon_t\rangle$, $B_j = \langle 0,p_j+t,q_j+t,4t,8t+j,5t\rangle \pmod n$, $j = 2,3,\ldots,t$;
$K_n \to G_{12}$: $B_1 = \langle t+p_1,t+q_1,0,3t+1-\epsilon_t,4t+1,8t+2\rangle$, $B_j = \langle t+p_j,t+q_j,0,3t+j,4t+1,8t+1+j\rangle \pmod n$, $j = 2,3,\ldots,t$, where $\{(p_r,q_r)\,|\,r = 1,2,\ldots,t\}$ is a t-system (see Definition 2.1).

(2) $n = 10t$, $K_n \to G_{11}$: B_j $(1 \le j \le t-1)$ the same as that in (1) $\pmod{n-1}$, $B_t = \langle 0,t+p_t,t+q_t,\infty,9t,5t\rangle \pmod{n-1}$.
$K_n \to G_{12}$: $B_j (1 \le j \le t-1)$ the same as that in (1) $\pmod{n-1}$, $B_t = \langle t+p_t,t+q_t,0,4t,4t+1,\infty\rangle \pmod{n-1}$, where $\{(p_r,q_r)\,|\,r = 1,2,3,\ldots,t\}$ is a t-system.

(3) $n = 10t+5$, $V(K_n) = \{y_x | y \in Z_{2t+1}, x \in I_5\}$,
$K_n \to G_{11}$: $B_{ij} = \langle (\frac{1}{2}j)_{i-1},j_i,0_i,0_{i-1},j_{i+1},(2j)_{i+4}\rangle$ if j even, $B_{ij} = \langle (t+\frac{1}{2}(j+1))_{i-1},j_i,0_i,0_{i-1},j_{i+1},(2j)_{i+4}\rangle$ otherwise, $\pmod{2t+1}$, $B_{5t+1} = \langle 0_1,0_2,0_3,1_3,1_5,1_2\rangle$ and $B_{5t+2} = \langle 0_1,0_5,0_4,1_3,1_4,1_2\rangle \pmod{2t+1}$;
$K_n \to G_{12}$: $B_{ij} = \langle (\frac{1}{2}j)_{i-1},j_i,0_i,j_{i+3},0_{i-1},j_{i+1}\rangle$ if j even, $B_{ij} = \langle (t+(j+1)/2)_{i-1},j_i,0_i,j_{i+3},0_{i-1},j_{i+1}\rangle$ otherwise, $B_{5t+1} = \langle 0_1,0_2,0_3,0_4,1_2,1_5\rangle$, $B_{5t+2} = \langle 0_1,0_5,0_4,0_2,1_3,1_5\rangle$, $\pmod{2t+1}$.

Remark: here $i = 1,2,\ldots,5$, $j = 1,2,\ldots,t$ and all subscripts are non-negative and are taken modulo 5.

(4) $n = 10t+6$, $V(K_n) = \{y_x | y \in Z_{2t+1}, x \in I_5\} \cup \{\infty\}$,
$K_n \to G_{11}$: B_{ij} $(1 \le i \le 5, 1 \le j \le t)$ the

same as that in (3)

$$B_{5t+1} = \langle 0_1, \infty, 0_5, 1_3, 1_5, 1_2 \rangle$$
$$B_{5t+2} = \langle 0_3, \infty, 0_4, 1_1, 1_4, 1_5 \rangle \quad (\text{mod } 2t+1)$$
$$B_{5t+3} = \langle 0_1, 0_2, 0_3, 1_4, 1_2, \infty \rangle$$

$K_n \to G_{12}$: $(1 \le i \le 5,\ 1 \le j \le t)$ the same as that in (3)

$$B_{5t+1} = \langle 0_1, 0_4, 0_3, \infty, 1_1, 1_2 \rangle$$
$$B_{5t+2} = \langle 0_1, 0_5, \infty, 0_2, 1_2, 1_3 \rangle \quad (\text{mod } 2t+1)$$
$$B_{5t+3} = \langle 0_2, 0_4, 0_5, 0_3, 1_4, \infty \rangle.$$

LEMMA 3.6. $K_6 \to G_{12}$.

PROOF. Let $V(K_6) = Z_6$. The subgraphs of the decomposition are $\langle 5,0,1,4,2,3 \rangle$, $\langle 4,0,2,5,3,1 \rangle$ and $\langle 4,5,3,0,2,1 \rangle$.

THEOREM 3.7. $n \in B(G_{11})$ iff $n \equiv 0,1 \pmod 5$, $n > 6$ and $n \in B(G_{12})$ iff $n \equiv 0,1 \pmod 5$.

PROOF. The conclusion follows from Lemmas 3.4, 3.5 and 3.6 and (1.1).

THEOREM 3.8. $n \in B(G_{13})$ iff $n \equiv 0,1 \pmod 5$.

PROOF. The necessity follows from (1.1). As for the sufficiency, we use Lemma 2.9 and 2.10. Let $V(K_{5,5}) = \{y_1 | y \in Z_5\} \cup \{y_2 | y \in Z_5\}$, $V(K_{5,4}) = (\{y_1 | y \in Z_4\} \cup \{\infty\}) \cup \{y_2 | y \in Z_4\}$. $K_{5,5}$ and $K_{5,4}$ can be decomposed into subgraphs isomorphic to G_{13}. The base graphs are given below.

$$K_{5,5} \to G_{13}: \langle 0_1, 0_2, 1_1, 3_2, 3_1, 4_2 \rangle \pmod 5$$
$$K_{5,4} \to G_{13}: \langle 0_1, 0_2, 1_1, 2_2, 3_2, \infty \rangle \pmod 4$$

Thus by Lemma 2.10, $K_{5,5m} \to G_{13}$ and $K_{5,5m+4} \to G_{13}$, where m is a positive integer. So, by Lemma 2.9 with $n_1 = 5$ and $n_2 = 5m$ or $5m+4$, if $K_{n_2+1} \to G_{13}$ and $K_6 \to G_{13}$ then $K_{n_2+6} \to G_{13}$. Thus in order to prove that $K_n \to G_{13}$ for $n \equiv 0,1 \pmod 5$ it is sufficient to prove that $K_{10} \to G_{13}$, $K_6 \to G_{13}$. This can easily be done:

$K_{10} \to G_{13}$: $\langle 0,1,5,2,3,\infty \rangle \pmod 9$;

$K_6 \to G_{13}$: $\langle 1,4,2,3,0,5 \rangle$, $\langle 0,1,5,2,3,4 \rangle$, $\langle 0,3,5,4,1,2 \rangle$, where $V(K_6) = Z_6$.

We will now consider the case $|E(G)| = 6$. The necessary condition can be derived easily from (1.1) and hence will be omitted in the proof.

THEOREM 3.9. $n \in B(G_i)$ iff $n \equiv 0,1,4,9 \pmod{12}$, for $i = 14,15,16,17$.

PROOF. The proof is similar to that of Theorem 3.8.

We need to construct $K_{n_i} \to G$ or $K_{n_1,n_2} \to G$ for some small values of n_i. They are exhibited as follows.

$K_{4,3} \to G_{14}$: $\langle 1,3,0,5,2,4\rangle\langle 0,4,1,6,2,3\rangle$;
$K_{4,3} \to G_{15}$: $\langle 0,3,2,6,4,1\rangle\langle 1,5,2,4,6,0\rangle$;
$K_{4,3} \to G_{16}$: $\langle 4,0,6,2,3,5\rangle\langle 3,0,5,1,4,6\rangle$;
$K_{2,6} \to G_{17}$: $\langle 2,1,4,0,7,6\rangle\langle 3,1,5,0,6,7\rangle$, where $V(K_{4,3}) = \{0,1,2\} \cup \{3,4,5,6\}$, $V(K_{2,6}) = \{0,1\} \cup \{2,3,\ldots,7\}$.

$$\left.\begin{array}{l} K_9 \to G_{14}: \langle 0_1,0_2,0_3,1_3,2_1,2_3\rangle,\ \langle 0_1,1_1,0_2,1_2,0_3,2_2\rangle \\ K_9 \to G_{15}: \langle 1_3,0_1,0_2,0_3,1_1,2_3\rangle,\ \langle 0_2,1_2,0_1,1_1,1_3,0_3\rangle \\ K_9 \to G_{16}: \langle 0_1,0_2,0_3,1_3,1_1,2_1\rangle,\ \langle 0_1,1_1,0_2,1_2,0_3,2_3\rangle \end{array}\right\} \pmod 3$$

$K_9 \to G_{17}$: $\langle 2_2,1_3,1_2,0_1,2_3,0_3\rangle$, $\langle 0_1,0_2,1_3,1_1,2_3,1_2\rangle$, where $V(K_9) = \{y_x \mid y \in Z_3,\ x \in I_3\}$.

$$\left.\begin{array}{l} K_{13} \to G_{14}: \langle 3,1,6,0,4,5\rangle \\ K_{13} \to G_{15}: \langle 0,2,1,6,4,5\rangle \\ K_{13} \to G_{16}: \langle 2,1,6,0,3,4\rangle \\ K_{13} \to G_{17}: \langle 6,0,3,1,2,4\rangle \end{array}\right\} \pmod{13}$$

$$\left.\begin{array}{l} K_{12} \to G_{14}: \langle 0,3,1,5,6,\infty\rangle \\ K_{12} \to G_{15}: \langle 0,2,1,5,3,\infty\rangle \\ K_{12} \to G_{16}: \langle 2,1,5,0,3,\infty\rangle \\ K_{12} \to G_{17}: \langle 2,1,5,0,3,\infty\rangle \end{array}\right\} \pmod{11}$$

Next, we assume that $V(K_{16}) = \{y_x \mid y \in Z_5,\ x \in I_3\}$.

$K_{16} \to G_{14}$: $\langle 2_1,1_3,2_2,0_2,1_2,1_1\rangle$ $\langle 2_2,1_1,2_3,0_3,1_3,\infty\rangle$
$\langle 2_3,1_2,2_1,0_1,\infty,0_2\rangle$ $\langle 2_1,0_3,3_2,1_1,1_3,1_2\rangle$ (mod 5)
$K_{16} \to G_{15}$: $\langle 2_1,1_3,2_2,0_2,\infty,1_2\rangle$ $\langle 2_2,1_1,2_3,0_3,3_2,1_2\rangle$
$\langle 2_3,1_2,2_1,0_1,3_3,\infty\rangle$ $\langle 2_1,0_3,3_2,1_1,2_3,\infty\rangle$ (mod 5)
$K_{16} \to G_{16}$: $\langle 2_1,1_3,2_2,0_2,1_2,\infty\rangle$ $\langle 2_2,1_1,2_3,0_3,1_3,0_2\rangle$
$\langle \infty,2_3,1_2,2_1,4_3,4_1\rangle$ $\langle 2_1,0_3,3_2,1_1,1_3,1_2\rangle$ (mod 5)
$K_{16} \to G_{17}$: $\langle 2_2,0_2,2_1,1_3,0_3,1_2\rangle$ $\langle 2_3,0_3,2_2,1_1,\infty,0_2\rangle$
$\langle 2_1,0_1,2_3,1_2,\infty,0_3\rangle$ $\langle 3_2,1_1,2_1,0_3,\infty,1_2\rangle$ (mod 5)

Now we can apply Lemmas 2.9 and 2.10 to obtain the required results, shown in Table 1.

TABLE 1

n	$12t+12, t\geq 1$, $12t+13, t\geq 1$	$12t+9, t\geq 1$	$12t+16, t\geq 1$
$K_n \to G_j$ (j=14,15,16,17) by Lemma 2.9 with $h = 2$, $n_2 = 12t$	$n_1=12, q=0$ and $q=1$	$n_1=8, q=1$	$n_1=16, q=0$
$K_{n_1,n_2} \to G_j$ (j=14,15,16) by Lemma 2.10 with $r_1=4$ $r_2=3$, $r_2'=3$, $b=4t$, $c=0$	a=3	a=2	a=4
$K_{n_1,n_2} \to G_{17}$ by Lemma 2.10, with $r_1=2$, $r_2=6$, $r_2'=6$, $b=2t$, $c=0$	a=6	a=4	a=8

EXPLANATION. $K_{12t+q} \to G_j$ ($q = 0$ or 1) is proved by using Lemma 2.9 recursively, which states that if $K_{12+q} \to G_j$ and $K_{12t+q} \to G_j$, then $K_{12(t+1)+q} \to G_j$.

THEOREM 3.10. For any $j = 18,19,20$, $n \in B(G_j)$ if and only if $n \equiv 0,1,4$ and $9 \pmod{12}$.

PROOF. For any $G \in \{G_{18},G_{19},G_{20}\}$, $K_{2,2,2}$, $K_{2,2,5}$ and K_m can be decomposed into subgraphs isomorphic to G, where $m = 9,12,13,16$ and 21. We indicate below.

$$K_{2,2,2} \to G_{18}: \langle 1,4,5,3,2,0\rangle\ \langle 1,3,0,4,2,5\rangle$$
$$K_{2,2,2} \to G_{19}: \langle 1,4,5,3,0,2\rangle\ \langle 2,3,0,4,5,1\rangle$$
$$K_{2,2,2} \to G_{20}: \langle 4,1,0,3,5,2\rangle\ \langle 3,2,0,4,5,1\rangle$$
$$K_{2,2,5} \to G_{18}: \langle 1,6,3,0,4,8\rangle\ \langle 3,5,2,0,1,8\rangle\ \langle 2,6,4,7,3,8\rangle\ \langle 5,4,1,7,2,8\rangle$$
$$K_{2,2,5} \to G_{19}: \langle 2,3,7,0,8,1\rangle\ \langle 1,3,6,5,0,4\rangle\ \langle 2,4,5,6,0,3\rangle\ \langle 1,4,8,0,7,2\rangle$$
$$K_{2,2,5} \to G_{20}: \langle 8,3,5,2,0,1\rangle\ \langle 8,2,6,4,0,3\rangle\ \langle 8,1,6,3,7,4\rangle\ \langle 8,4,5,1,7,2\rangle$$

Here $V(K_{2,2,2}) = \{1,2\} \cup \{3,4\} \cup \{5,0\}$ and $V(K_{2,2,5}) = \{1,2\} \cup \{3,4\} \cup \{5,6,7,8,0\}$.

$$\left.\begin{array}{l} K_9 \to G_{18}: \langle 0_3,1_3,0_1,2_2,2_3,1_2\rangle\langle 0_1,0_2,1_2,0_3,1_1,2_1\rangle \\ K_9 \to G_{19}: \langle 0_3,1_3,0_1,1_2,2_1,1_1\rangle\langle 0_1,0_2,1_2,2_2,0_3,2_3\rangle \\ K_9 \to G_{20}: \langle 1_1,0_3,1_3,0_1,2_2,2_3\rangle\langle 1_1,0_1,0_2,1_2,0_3,2_2\rangle \end{array}\right\}\ (\text{mod } 3)$$

Here $V(K_q) = \{y_x | y \in z_3,\ x \in I_3\}$

$$\left.\begin{array}{l} K_{12} \to G_{18}: \langle 1,0,5,7,10,\infty\rangle \\ K_{12} \to G_{19}: \langle 1,0,5,4,\infty,7\rangle \\ K_{12} \to G_{20}: \langle \infty,1,0,5,8,10\rangle \end{array}\right\}\ (\text{mod } 11)$$

$$\left.\begin{array}{l} K_{13} \to G_{18}: \langle 1,0,5,7,10,4\rangle \\ K_{13} \to G_{19}: \langle 1,0,5,4,6,7\rangle \\ K_{13} \to G_{20}: \langle 7,1,0,5,8,10\rangle \end{array}\right\}\ (\text{mod } 13)$$

$$K_{16} \to G_{18}: \langle 0_3,1_3,0_1,2_2,2_3,\infty\rangle\langle 2_3,0_3,3_1,2_2,4_1,2_1\rangle\ \langle 0_1,0_2,1_2,4_3,3_2,\infty\rangle\langle 0_2,2_2,4_3,1_1,2_1,\infty\rangle\ (\text{mod } 5)$$
$$K_{16} \to G_{19}: \langle 0_3,1_3,0_1,2_2,\infty,1_1\rangle\langle 2_3,0_3,3_1,1_2,2_1,1_1\rangle\ \langle 0_1,0_2,1_2,\infty,1_1,3_1\rangle\langle 0_2,2_2,4_3,\infty,0_1,4_2\rangle\ (\text{mod } 5)$$
$$K_{16} \to G_{20}: \langle 0_2,0_3,1_3,0_1,\infty,1_2\rangle\langle \infty,2_3,0_3,3_1,2_2,4_1\rangle\ \langle 2_1,0_1,0_2,1_2,4_3,3_2\rangle\langle 3_1,0_2,2_2,4_3,1_1,2_1\rangle\ (\text{mod } 5)$$
$(V(K_{16}) = \{y_x | y \in Z_5,\ x \in I_3\} \cup \{\infty\})$

$$K_{21} \to G_{18}: \langle 1_3,0_3,3_1,6_1,6_3,3_3\rangle\langle 6_1,2_3,0_3,1_1,3_2,6_3\rangle\ \langle 2_2,0_2,3_1,4_1,4_2,1_3\rangle\langle 1_2,0_2,6_3,4_1,2_2,2_3\rangle\ \langle 2_1,0_1,3_2,4_3,2_2,5_2\rangle\ (\text{mod } 7)$$

$K_{21} \to G_{19}$: $\langle 1_3,0_3,3_1,1_1,0_2,2_3\rangle\langle 6_1,2_3,0_3,6_2,5_2,4_2\rangle$
$\langle 2_2,0_2,3_1,4_3,1_3,5_3\rangle\langle 1_2,0_2,6_3,3_1,5_1,3_3\rangle$
$\langle 2_1,0_1,3_2,5_1,1_1,6_2\rangle$ (mod 7)

$K_{21} \to G_{20}$: $\langle 4_3,1_3,0_3,3_1,6_1,6_3\rangle\langle 1_3,6_1,2_3,0_3,1_1,3_2\rangle$
$\langle 4_1,2_2,0_2,3_1,3_2,0_3\rangle\langle 4_2,1_2,0_2,6_3,3_2,3_3\rangle$
$\langle 3_1,2_1,0_1,3_2,4_3,2_2\rangle$ (mod 7)
$(V(K_{21}) = \{y_x | y \in Z_7,\ x \in I_3\})$

$K_{24} \to G_{18}$: $\langle 0,11,4,9,12,6\rangle\langle 0,10,2,3,12,\infty\rangle$ (mod 23)
$K_{24} \to G_{19}$: $\langle 0,11,4,6,\infty,9\rangle$ $\langle 0,10,2,9,13,3\rangle$ (mod 23)
$K_{24} \to G_{20}$: $\langle 6,0,11,4,9,12\rangle\langle \infty,0,10,2,3,12\rangle$ (mod 23)

Now we can apply Lemmas 2.8 and 2.9 to get the required decomposition. The proof is given in Table 2.

TABLE 2

n	$K_n \to G_j$ (j = 18,19,20) by Lemma 2.9 with	$K_{n_1,n_2,\ldots,n_h} \to G_j$ (j=18,19,20) by Lemma 2.8 with k = 3
12t+21 $t \geq 2$	$h=t+1, q=0, n_{t+1}=21$, $n_1=n_2=\ldots=n_t=12$	$m=6$, $u=t+1$, $r=2$, $s_1=2$, $s_2=5$, $a=3$, $b=3$
12t+16 $t \geq 2$	$h=t+1, q=1, n_{t+1}=15$, $n_1=n_2=\ldots=n_t=12$	$m=6$, $u=t+1$, $r=2$, $s_1=2$, $s_2=5$, $a=5$, $b=1$
12t+12, or 12t+13 $t \geq 2$	$h=t+1$, $q=0$ or 1, $n_1=\ldots=n_h=12$	$m=6$, $u=t+1$, $r=2$, $s_1=2$, $s_2=5$, $a=6$, $b=0$
25	$h=3, q=1, n_1=n_2=n_3=8$	$m=4$, $u=3$, $s_1=s_2=a=b=r=2$
28	$h=3, q=1, n_1=n_2=8$, $n_3=11$	$m=4$, $u=3$, $s_1=2$, $s_2=5$, $a=3$, $b=1$, $r=2$
33	$h=4, q=1, n_1=\ldots=n_4=8$	$m=u=4$, $s_1=s_2=a=b=r=2$

THEOREM 3.11. A necessary and sufficient condition for the existence of $K_n \to G_{21}$ is $n \equiv 1,9 \pmod{12}$ and $n > 9$.

PROOF. Consider the graph K_{12t+9}. It is clear that

$K_{12t+9} \to G_{21}$ is equivalent to the existence of a (12t+9,3,1)-BIBD in which the block family can be partitioned into pairs of disjoint blocks. By Lemma 2.6, there exists a (12t+9,3,1)-RBIBD. For 12t+9 > 9, we can take a block from every parallel class of the RBIBD to get 3t+2 pairs of disjoint blocks and use the remaining blocks of every parallel class to yield (6t+4)(2t+1) pairs. Thus, when 12t+9 > 9 we have $K_{12t+9} \to G_{21}$. However, $K_9 \not\to G_{21}$. This is because there is only one (9,3,1)-BIBD up to isomorphism (see [10]), i.e., (9,3,1)-RBIBD. In fact, any two disjoint blocks are contained in a parallel class and the third block of which intersects all the remaining nine blocks. Now we construct $K_{12t+1} \to G_{21}$. Let $\{(p_i,q_i) \mid 1 \le i \le 2t\}$ be a (2t)-system. The subgraphs of the decomposition $K_{12t+1} \to G_{21}$ are

$$\langle 0,p_i+2t,q_i+2t,2t,p_{t+i}+4t,q_{t+i}+4t\rangle \pmod{12t+1},$$

where i = 1,2,...,t. Thus the proof is completed.

THEOREM 3.12. For i = 22,23,24, $n \in B(G_i)$ iff $n \equiv 0,1,4,9$ (mod 12).

PROOF. As in Theorem 3.10, we first give the following decompositions which are used in the recursive construction.

Let $V(K_{4,4,4}) = \bigcup_{i=1}^{3} \{y_i \mid y \in Z_4\}$ and $V(K_{4,4,7}) = (\bigcup_{i=1}^{2} \{y_i \mid y \in Z_4\}) \cup \{y_3,\infty_j \mid j = 1,2,3,\ y \in Z_4\}$.

$$\left.\begin{array}{l} K_{4,4,4} \to G_{22}: \langle 0_3,0_2,1_1,2_3,2_2,3_3\rangle\langle 3_3,3_1,1_2,1_1,0_3,2_3\rangle \\ K_{4.4.4} \to G_{23}: \langle 0_3,0_2,1_1,2_2,2_3,0_1\rangle\langle 3_3,3_1,1_2,0_3,2_3,1_3\rangle \\ K_{4,4,4} \to G_{24}: \langle 3_3,3_1,1_2,0_3,2_3,0_1\rangle\langle 0_3,0_2,1_1,2_3,2_2,2_1\rangle \end{array}\right\} \pmod 4$$

$$K_{4,4,7} \to G_{22}: \begin{array}{l} \langle 1_2,\infty_1,0_1,\infty_3,1_3,3_3\rangle \\ \langle 1_1,\infty_2,0_2,\infty_3,0_1,1_3\rangle \\ \langle 2_1,0_2,0_3,0_1,2_2,1_2\rangle \pmod 4 \end{array}$$

$$K_{4,4,7} \to G_{23}: \begin{array}{l} \langle \infty_1,1_2,0_1,\infty_3,1_3,3_3\rangle \\ \langle \infty_2,1_1,0_2,\infty_3,1_3,1_2\rangle \\ \langle 0_2,2_1,0_3,0_1,1_2,1_3\rangle \pmod 4 \end{array}$$

$$K_{4,4,7} \to G_{24}: \langle \infty_1,1_2,0_1,\infty_3,1_3,2_1\rangle$$
$$\langle \infty_2,1_1,0_2,\infty_3,1_3,3_2\rangle$$
$$\langle 0_2,2_1,0_3,0_1,1_2,1_1\rangle \pmod 4$$

$$\left.\begin{array}{l} K_9 \to G_{22}: \langle 1_2,0_3,0_1,1_1,1_3,2_3\rangle\langle 0_3,1_3,0_2,0_1,1_1,1_2\rangle \\ K_9 \to G_{23}: \langle 0_3,1_2,0_1,1_1,1_3,2_2\rangle\langle 0_3,1_3,0_2,0_1,1_1,2_1\rangle \\ K_9 \to G_{24}: \langle 0_3,1_2,0_1,1_1,1_3,2_1\rangle\langle 0_3,1_3,0_2,0_1,1_2,2_1\rangle \end{array}\right\} \pmod 3$$

$$(V(K_9) = \{y_x \mid y \in Z_3,\ x \in I_3\})$$

$$\left.\begin{array}{l} K_{12} \to G_{22}: \langle 1,4,0,2,5,\infty\rangle \\ K_{12} \to G_{23}: \langle 1,4,0,2,5,\infty\rangle \\ K_{12} \to G_{24}: \langle 1,4,0,\infty,2,7\rangle \end{array}\right\} \pmod{11}$$

$$\left.\begin{array}{l} K_{13} \to G_{22}: \langle 4,1,0,2,5,6\rangle \\ K_{13} \to G_{23}: \langle 4,1,0,2,5,7\rangle \\ K_{13} \to G_{24}: \langle 4,1,0,2,5,12\rangle \end{array}\right\} \pmod{13}$$

$$K_{16} \to G_{22}: \langle 2_1,2_2,0_1,2_3,3_3,\infty\rangle\langle 1_3,2_3,0_2,1_2,2_2,\infty\rangle$$
$$\langle 2_2,2_3,0_3,1_1,1_2,\infty\rangle\langle 1_1,1_3,0_1,1_2,3_2,4_2\rangle \pmod 5$$
$$K_{16} \to G_{23}: \langle 2_2,2_1,0_1,2_3,\infty,0_3\rangle\langle 2_3,1_3,0_2,2_2,\infty,2_1\rangle$$
$$\langle 2_3,2_2,0_3,1_2,\infty,3_2\rangle\langle 1_3,1_1,0_1,3_2,4_2,2_2\rangle \pmod 5$$
$$K_{16} \to G_{24}: \langle 2_1,2_2,0_1,\infty,2_3,4_1\rangle\langle 1_3,2_3,0_2,\infty,1_2,3_2\rangle$$
$$\langle 2_2,2_3,0_3,\infty,1_1,0_2\rangle\langle 1_1,1_3,0_1,3_2,1_2,0_3\rangle \pmod 5$$
$$(V(K_{16}) = \{y_x \mid y \in Z_5,\ x \in I_3\})$$

$$K_{21} \to G_{22}: \langle 1_2,3_1,0_1,1_1,2_1,2_2\rangle\langle 3_2,4_1,0_2,2_2,3_1,4_3\rangle$$
$$\langle 1_1,3_3,0_3,1_3,2_1,2_3\rangle\langle 0_2,0_3,0_1,1_3,3_3,4_3\rangle$$
$$\langle 1_2,6_3,0_2,1_3,2_3,3_3\rangle \pmod 7$$
$$K_{21} \to G_{23}: \langle 3_1,1_2,0_1,1_1,2_1,6_1\rangle\langle 3_2,4_1,0_2,2_2,4_3,1_2\rangle$$
$$\langle 3_3,1_1,0_3,1_3,2_3,6_3\rangle\langle 0_2,0_3,0_1,1_3,3_3,3_1\rangle$$
$$\langle 1_2,6_3,0_2,1_3,3_3,4_2\rangle \pmod 7$$
$$K_{21} \to G_{24}: \langle 1_2,3_1,0_1,1_1,2_1,4_2\rangle\langle 3_2,4_1,0_2,3_1,2_2,6_3\rangle$$
$$\langle 1_1,3_3,0_3,2_1,1_3,6_3\rangle\langle 0_2,0_3,0_1,4_3,1_3,5_1\rangle$$
$$\langle 1_2,6_3,0_2,2_3,3_3,2_2\rangle \pmod 7$$
$$(V(K_{21}) = \{y_x \mid y \in Z_7,\ x \in I_3\})$$

$$\left.\begin{array}{l} K_{24} \to G_{22}: \langle 1,4,0,2,5,\infty\rangle\langle 6,13,0,8,9,11\rangle \\ K_{24} \to G_{23}: \langle 1,4,0,2,5,\infty\rangle\langle 6,13,0,8,9,2\rangle \\ K_{24} \to G_{24}: \langle 1,4,0,\infty,2,7\rangle\langle 6,13,0,11,8,17\rangle \end{array}\right\} \pmod{23}$$

Next, we use Lemma 2.8 and 2.9 in Table 3.

TABLE 3

n	$K_n \to G_j$ (j=22,23,24) by Lemma 2.9 with	$K_{n_1,n_2\ldots n_h} \to G_j$ by Lemma 2.8 with $s_1=4$, $s_2=7$, r=4, k=3
24t+9 $t\geq 1$	h = 3t+1, q = 1, $n_1=n_2=\ldots=n_h=8$	m = 2, u = 3t+1, a = 2, b = 0
24t+21 $t\geq 1$	h = 2t+1, q = 0, $n_{2t+1} = 21$, $n_1=n_2=\ldots=n_{2t}=12$	m = 3, u = 2t+1, a = 0, b = 3
24t+12 $t\geq 1$	h = 2t+1, q = 0, $n_1=n_2=\ldots=n_h=12$	m = 3, u = 2t+1, a = 3, b = 0
24t+24 $t\geq 1$	h = t+1, q = 0, $n_1=n_2=\ldots=n_h=24$; h = 3, q = 0, $n_1 = n_2 = n_3 = 16$	m = 6, u = t+1, a = 6, b = 0; m = 4, u = 3, a = 4, b = 0
24t+13 $t\geq 1$	h = 2t+1, q = 1, $n_1 = n_2 = \ldots = n_h = 12$	m = 3, u = 2t+1, a = 3, b = 0
24t+25 $t\geq 0$	h = 3t+3, q = 1, $n_1 = n_2 = \ldots = n_h = 8$	m = 2, u = 3t+3, a = 2, b = 0
24t+28 $t\geq 0$	h = 3t+3, q = 1, $n_{3t+3} = 11$, $n_1=n_2=\ldots=n_{3t+2}=8$	m = 2, u = 3t+3, a = 1, b = 1
24t+16 $t\geq 1$	h = 2t+1, q = 1, $n_h = 15$, $n_1=n_2=\ldots=n_{2t}=12$	m = 3, u = 2t+1, a = 2, b = 1

THEOREM 3.13. A necessary and sufficient condition for the existence of $K_n \to G_{25}$ is $n \equiv 0,1,4,9 \pmod{12}$.

PROOF. Similar to Theorem 3.12. We only construct the following decomposition. Take $V(K_{4,4,4}) = \{0,1,2,3\} \cap \{4,5,6,7\} \cap \{8,9,10,11\}$ and take $V(H)$ as in Theorem 3.12 for any $H \in \{K_{4,4,7}, K_i \mid i = 9,12,13,16,21,24\}$.

$$K_{4,4,4} \to G_{25}: \langle 6,1,8,0,2,5\rangle\langle 7,0,9,1,11,6\rangle\langle 6,2,10,3,1,4\rangle$$
$$\langle 7,3,11,2,9,6\rangle\langle 2,8,4,9,0,5\rangle\langle 3,9,5,8,10,7\rangle$$
$$\langle 0,10,4,11,8,7\rangle\langle 1,11,5,10,3,4\rangle$$
$$K_{4,4,7} \to G_{25}: \langle 0_1,1_2,0_3,0_2,\infty_1,1_1\rangle\langle 0_1,3_3,2_2,\infty_2,\infty_1,1_2\rangle$$
$$\langle 3_2,1_3,0_1,\infty_3,2_1,0_3\rangle \pmod 4$$
$$K_9 \to G_{25}: \langle 2_3,2_1,1_2,1_3,2_2,0_2\rangle\langle 0_1,2_3,1_1,1_2,1_3,2_2\rangle \pmod 3$$
$$K_{12} \to G_{25}: \langle 1,3,0,5,2,\infty\rangle \pmod{11}$$
$$K_{13} \to G_{25}: \langle 1,3,0,5,2,8\rangle \pmod{13}$$
$$K_{16} \to G_{25}: \langle 0_3,0_2,1_3,0_1,4_1,2_3\rangle\langle 0_3,\infty,3_1,2_3,3_2,1_3\rangle$$
$$\langle 0_1,2_1,0_2,1_2,\infty,3_2\rangle\langle 0_2,3_1,2_2,4_3,0_1,1_1\rangle \pmod 5$$
$$K_{21} \to G_{25}: \langle 1_3,1_1,0_3,3_1,4_1,2_2\rangle\langle 6_1,0_1,2_3,0_3,0_2,1_3\rangle$$
$$\langle 2_2,2_3,0_2,3_1,1_1,4_1\rangle\langle 1_2,4_3,0_2,6_3,2_2,5_2\rangle$$
$$\langle 2_1,2_2,0_1,3_2,0_3,3_3\rangle \pmod 7$$
$$K_{24} \to G_{25}: \langle 0,5,11,4,1,\infty\rangle\langle 0,1,10,2,3,6\rangle \pmod{23}$$

So the proof is complete.

THEOREM 3.14. For $j = 26,27$, $n \in B(G_j)$ if and only if $n \equiv 0,1,4,9 \pmod{12}$.

PROOF. Using the method of differences, we can construct the following decomposition $H \to G_j$, where $V(H)$ is the same as in Theorem 3.12.

$$K_{4,4,4} \to G_{26}: \langle 3_2,3_3,0_1,0_2,3_1,1_3\rangle$$
$$\langle 0_3,2_2,0_1,2_3,1_1,3_2\rangle \pmod 4$$
$$K_{4,4,4} \to G_{27}: \langle 0_3,2_1,2_2,1_1,3_2,1_3\rangle$$
$$\langle 2_2,3_1,2_3,1_1,1_3,1_2\rangle \pmod 4$$

$$K_{4,4,7} \to G_{26}: \langle \infty_1,2_2,0_1,0_3,1_1,2_1\rangle\langle \infty_2,1_1,0_2,0_3,3_1,1_2\rangle$$
$$\langle \infty_3,1_2,0_1,0_2,1_3,2_3\rangle \pmod 4$$

$K_{4,4,7} \to G_{27}$: $\langle 3_2,\infty_1,2_1,2_2,0_3,0_1\rangle\langle 2_3,1_1,\infty_2,2_2,3_1,0_2\rangle$
$\langle 2_2,\infty_3,1_1,3_2,2_3,0_1\rangle$ (mod 4)

$K_9 \to G_{26}$: $\langle 1_1,1_3,0_1,2_2,0_3,2_3\rangle\langle 0_1,0_2,1_2,0_3,1_3,1_1\rangle$ (mod 3)
$K_9 \to G_{27}$: $\langle 2_2,0_2,2_3,1_2,0_1,1_3\rangle\langle 1_1,1_2,1_3,0_3,0_1,2_3\rangle$ (mod 3)

$K_{12} \to G_{26}$: $\langle 4,1,0,5,\infty,7\rangle$
$K_{12} \to G_{27}$: $\langle 8,4,6,1,0,\infty\rangle$ } (mod 11)

$K_{13} \to G_{26}$: $\langle 4,1,0,5,11,7\rangle$
$K_{13} \to G_{27}$: $\langle 8,4,7,1,0,2\rangle$ } (mod 13)

$K_{16} \to G_{26}$: $\langle 3_2,2_2,0_1,1_1,0_3,2_3\rangle\langle 4_2,1_2,0_1,2_1,2_2,4_3\rangle$
$\langle 3_3,2_3,0_2,0_3,0_1,2_1\rangle\langle 4_3,1_3,0_2,\infty,0_3,0_1\rangle$ (mod 5)
$K_{16} \to G_{27}$: $\langle 1_2,0_3,4_2,4_1,0_1,2_3\rangle\langle 2_1,1_2,3_3,4_1,2_3,1_3\rangle$
$\langle \infty,0_1,2_2,4_2,2_3,0_3\rangle\langle 2_1,0_1,1_3,1_2,0_2,\infty\rangle$ (mod 5)

$K_{21} \to G_{26}$: $\langle 2_1,3_1,0_3,4_1,3_2,5_2\rangle\langle 1_3,4_3,0_2,5_3,3_1,4_1\rangle$
$\langle 2_3,2_1,0_2,3_1,0_1,1_1\rangle\langle 1_2,4_2,0_3,0_2,2_2,0_1\rangle$
$\langle 2_2,3_2,0_1,6_3,4_3,5_3\rangle$ (mod 7)
$K_{21} \to G_{27}$: $\langle 4_2,4_1,0_1,1_2,0_3,1_3\rangle\langle 3_3,4_1,2_3,2_1,1_2,4_2\rangle$
$\langle 0_1,2_2,4_2,2_3,4_3,0_2\rangle\langle 1_3,1_2,0_2,2_1,0_1,4_2\rangle$
$\langle 0_1,2_3,5_3,4_2,1_1,4_3\rangle$ (mod 7)

$K_{24} \to G_{26}$: $\langle 8,2,0,4,13,\infty\rangle\langle 11,10,0,5,8,12\rangle$
$K_{24} \to G_{27}$: $\langle 1,3,7,10,0,9\rangle\langle 5,11,4,12,0,\infty\rangle$ } (mod 23)

The remaining verification is quite similar to that of Theorem 3.12 and hence omitted.

The following result is contained in [2].

THEOREM 3.15. $n \in B(G_{28})$ if and only if $n \equiv 1$ or 9 (mod 12).

Let us conclude with the remark that we have shown the necessary condition (1.1) for the existence of $K_n \to G$ is also sufficient for 28 non-isomorphic graphs in Figure 1, except for $K_6 \not\to G_j$ (j = 7,9,10,11) and $K_9 \not\to G_{21}$. The conditions for the existence of G-designs on $|V(G)| = 6$ and $|E(G)| > 6$ remain to be found; this is now under investigation.

REFERENCES

[1] J.-C. Bermond and J. Schönheim, G-decomposition of K_n, where G has four vertices or less, Discrete Math. 19(1977), 113-120.

[2] J.-C. Bermond and D. Sotteau, Graph decompositions and G-designs, Congressus Numerantium 15(1976), 53-72.

[3] J.-C. Bermond, C. Huang, A. Rosa, and D. Sotteau, Decomposition of complete graphs into isomorphic subgraphs with five vertices, Ars Combinatoria 10(1980), 293-318.

[4] M. Hall, Jr., Combinatorial Theory, Blaisdell, Waltham, Mass. 1967.

[5] H. Hanani, Balanced incomplete block designs and related designs, Discrete Math. 11(1975), 255-369.

[6] H. Hanani, D.K. Ray-Chaudhuri and R.M. Wilson, On resolvable designs, Discrete Math. 3(1972), 343-357.

[7] F. Harary, Graph Theory, Addison-Wesley, Reading, Mass. 1966.

[8] C. Huang, Balanced graph designs on small graphs, Utilitas Math. 10(1976), 77-108.

[9] C. Huang and A. Rosa, Decomposition of complete graphs into trees, Ars Combinatoria 5(1978), 23-63.

[10] R. Mathon and A. Rosa, Tables of parameters of BIBD with $r \leq 41$ including existence, enumeration, and resolvability results, Ann. Discrete Math. 26(1985), 275-308.

Existence of Balanced Incomplete Block Designs with $k = 7$ and $\lambda = 1$

ZHANG Yusen, Dalian University of Technology, Dalian, China

1. INTRODUCTION

Let X be a v-set. Let A be a family of some k-subsets (called blocks) of X. The pair (X,A) is called a balanced incomplete block design, denoted by (v,k,λ)-BIBD, if any two distinct elements of X are contained in exactly λ blocks of A. It is known that for any given integers v, k and λ, a (v,k,λ)-BIBD exists only if

$$\begin{aligned} \lambda(v-1) &\equiv 0 \pmod{k-1} \\ \lambda v(v-1) &\equiv 0 \pmod{k(k-1)}. \end{aligned} \tag{1.1}$$

When $k = 7$ and $\lambda = 1$, the necessary condition (1.1) is equivalent to

$$v \equiv 1,7 \pmod{42}. \tag{1.2}$$

In this paper it is proved that the condition is also sufficient for $v \geq 343687$:

THEOREM 1.1. If $v \geq 343687$ and $v \equiv 1,7 \pmod{42}$, then there exists a $(v,7,1)$-BIBD.

We use the following notations:

$$B(k) = \{v: \text{a } (v,k,1)\text{-BIBD exists}\}$$
$$R(k) = \{r: (k-1)r+1 \in B(k)\}$$
$$U(k) = \{u: ku+1 \in R(k)\}$$
$$V(k) = \{v: kv \in R(k)\}.$$

2. PRELIMINARIES

In the following sections we use the ideas of pairwise balanced design (PBD), incomplete PBD (IPBD), group divisible design (GDD), transversal design (TD) and PBD-closure. For definitions and notations, we refer to [1]. Let $(v,K \cup \{e^*\},1)$-PBD denote a PBD which has exactly one block of size e and other block sizes in K, $F_K[u]$ denote the set of positive integers v for which there exists a $(v,K,1)$-PBD with a flat of order u, and $F_K[0] = \{0\} \cup B(K)$. It is well known that a TD(k,n) is equivalent to k-2 mutually orthogonal latin squares (MOLS) of order n. For a list of lower bounds on the number of MOLS of all orders up to 100000, we refer the reader to Brouwer [2].

$$P_k(n) = \{v: \text{a } TD(k+2,v)-TD(k+2,n)\}$$

We need the following results:

LEMMA 2.1 [8]. Let $m,d \geq 0$ be such that $m+d \in F_{\{k\}}[d]$ and there exists a $TD(k,m/(k-1))$. Then, for every $v \in B(k)$,

$$\left(\frac{m}{k-1}\right)(v-1)+d \in B(k).$$

LEMMA 2.2 (Filling in Groups). Suppose $(X,\mathcal{G},A)$ is a P-GDD, where P is a PBD-closed set. If $|G| \in P$ for all $G \in \mathcal{G}$ then $|X| \in P$. If $|G|+1 \in P$ for all $G \in \mathcal{G}$, then $|X|+1 \in P$.

LEMMA 2.3 (Singular Indirect Product). Suppose K is a set of positive integers and $u \in K$; suppose v, w and a are integers such that $0 \leq a \leq w \leq v$; and suppose that the following designs exist:

1) a $TD(u,v-a)-TD(u,w-a)$
2) a (v,w,K)-IPBD, and
3) a $(u(w-a)+a,K)$-PBD.

Then there is a $(u(v-a)+a,K)$-PBD, that contains flats of order u and $u(w-a)+a$. Hence, in particular, $u(v-a)+a \in B(K)$.

LEMMA 2.4. Suppose there exist K-GDDs of group-type 1^{n+1} and $1^n c^1$. If there exists a TD(n+1,m) and $0 \leq t \leq m$, then there exists a K-GDD of group-type $m^n(tc-t+m)^1$.

LEMMA 2.5. Suppose there exists a TD(k,n), a TD(k,n-1) and a TD(n+1,m) and $0 \leq t \leq m$. Then there is a {k,n}-GDD having group-type $(mk-m)^{n-1}(mn-m)^1(tk-t)^1$.

LEMMA 2.6. Suppose there exists a TD(n,m), and a TD(n,n). There is a {n}-GDD having group-type $(mn-m)^{n+1}$.

LEMMA 2.7. If $v \in B(K)$, and $K \subset B(k,\lambda)$, then $v \in B(k,\lambda)$ holds.

LEMMA 2.8. If there exists a TD(s,r) and if $\{s,r+1\} \subset B(K,\lambda)$, then, $rs+1 \in B(K,\lambda)$.

LEMMA 2.9. If there exists a TD(s,r) and if $\{s,r\} \subset B(K,\lambda)$, then $rs \in B(K,\lambda)$.

LEMMA 2.10. If $m+k \in B(k)$, then $GD(k,\lambda,m)+k \in B(k,\lambda)$.

In order to apply the Singular Indirect Product, we need incomplete transversal designs [1]. We use constructions given in [1], [5] and [4] to produce these.

LEMMA 2.11. Suppose there exists a TD(k,m), a TD(k,m+1) and a TD(k+1,t), and suppose $0 \leq u \leq t$. Then there exists a TD(k,mt+u)-TD(k,u).

LEMMA 2.12. Suppose there exists a TD(k,m), a TD(k,m+1), a TD(k,m+2), a TD(k+2,t) and a TD(k,u), and $0 \leq u, v \leq t$. Then there exists a TD(k,mt+u+v)-TD(k,v).

We also need the following:

LEMMA 2.13. The following conditions are pairwise equivalent:

1) a (v,k,1)-RBIBD exists;

2) a $v+r \in B(k+1,r^*)$, where $r = (v-1)/(k-1)$;

3) There exists a $GD[\{k+1,r^*\},1,k; kr]$ with group-type k^r.

When $k = 6$, this becomes

LEMMA 2.14. The following conditions are pairwise equivalent:

1) $v \in RB(6)$;

2) $v+r \in B(7,r^*)$, where $r = (v-1)/5$;

3) There exists a $GD[\{7,r^*\},1,6; 6r]$ with group-type 6^r.

3. SOME APPLICATIONS OF THESE RESULTS

In the following, we will often omit the verification that a design is a TD(k,n). The reader can refer to [2].

In this section, we give some results that we need.

LEMMA 3.1. $\{1,8,15,28,36,56,70,77,147,287,511,553,637,651,770,826\} \subset R(7)$.

PROOF. Obviously, $7 \in B(7)$. From [9], we know $\{49,91,169,217\} \subset B(7)$, and from [14], we can obtain $\{337,421,463,883,1723,3067,3319,3823,3907,4621,4957\} \subset B(7)$. Then the conclusion follows.

LEMMA 3.2. If there exists a TD(7,m), then $48m+1 \in B(7,6m+1)$.

PROOF. Taking $n = 7$ in Lemma 2.6, noting that a TD(7,7) exists, construct a $\{7\}$-GDD of group-type $(6m)^8$, then fill in groups.

COROLLARY 3.3. $\{64,448,560,616,1176,2296\} \subset R(7)$.

PROOF. Set $m = 8,56,70,77,147,287$ in Lemma 3.2. We know $385 \in B(7,49)$, $2688 \in B(7,337)$, $3364 \in B(7,421)$, $3697 \in B(7,463)$, $7057 \in B(7,883)$, $13777 \in B(7,1723)$. Then the conclusion follows from Lemma 3.1 and Lemma 2.7.

LEMMA 3.4. If $r \in R(8)$, then $7r+1 \in R(7)$.

PROOF. If $r \in R(8)$, then $7r+1 \in B(8)$, so we have $6(7r+1)+1 \in$

B(7) (see [3], Lemma 2.17), so $7r+1 \in R(7)$.

COROLLARY 3.5. $\{120, 344,904,1072,1912,2024,2409,2416,2472, 2528,2752,2801,2864,2920,3032,3200,3312,4320\} \subset R(7)$.

PROOF. Since $\{17,49,129,153,273,289,344,345,353,361,393,400, 409,417,433,457,473,617\} \subset R(8)$ from [13], we have our conclusion.

LEMMA 3.6. $\{232,757\} \subset R(7)$.

PROOF. From [9], we have $232 \in B(8)$, $757 \in B(8)$, and from the proof of Lemma 3.4 we know $B(8) \subset R(7)$. Hence $232 \in R(7)$ and $757 \in R(7)$.

LEMMA 3.7. If $\{r,s\} \subset R(k)$ and there exists a TD(k,r), then $rs \in R(k)$.

PROOF. The conclusion follows from Lemma 2.1.

COROLLARY 3.8. $\{224,288,1856,2296\} \subset R(7)$.

PROOF. Take $k = 7$, $r = 8$ and $s = 28$, 36, 232, 287 in Lemma 3.7. Then the result follows from Lemma 3.1 and Lemma 3.6.

LEMMA 3.9. $\{57,106,197,253,393,400,491,540,694,736,743,841, 1030,2010,3578,3872\} \subset R(7)$.

PROOF. Take $r = 49,91,169,217,337,343,421,463,595,631,637,721, 883,1723,3067$ and 3319, $s = 7$, $\lambda = 1$ and $K = \{7\}$ in Lemma 2.9, noting that there exists a TD(7,r). Then the conclusion follows from Lemma 3.1, Corollary 3.3, Corollary 3.5, Corollary 3.11 and Corollary 3.16.

LEMMA 3.10. Suppose there exists a TD(7,m), and $m+1 \in B(7)$. Then $7m+1 \in B(7)$.

PROOF. This follows from Lemma 2.8.

COROLLARY 3.11. $\{105,196,252,392,399,448,490,539,735,742,840, 1029,1085,1134,1330,1372,1379,1568,2744,2800\} \subset R(7)$.

PROOF. The result follows from Lemma 3.10 and Table 1.

TABLE 1

m	7m+1	R(7)	check condition
90	631	105	15 ∈ R(7)
168	1177	196	28 ∈ R(7)
216	1513	252	36 ∈ R(7)
336	2353	392	56 ∈ R(7)
342	2395	399	57 ∈ R(7)
384	2689	448	64 ∈ R(7)
420	2941	490	70 ∈ R(7)
462	3235	539	77 ∈ R(7)
630	4411	735	105 ∈ R(7)
636	4453	742	106 ∈ R(7)
720	5041	840	120 ∈ R(7)
882	6175	1029	147 ∈ R(7)
930	6511	1085	155 ∈ R(7)
972	6805	1134	162 ∈ R(7)
1140	7981	1330	190 ∈ R(7)
1176	8233	1372	196 ∈ R(7)
1182	8275	1379	197 ∈ R(7)
1344	9409	1568	224 ∈ R(7)
2352	16465	2744	392 ∈ R(7)
2400	16801	2800	400 ∈ R(7)

LEMMA 3.12. $\{126,156,756,936,4536,5616\} \subset RB(6)$.

PROOF. It has been shown in [10] that $q^3+1 \in RB(q+1)$ if q is a prime power. Take q = 5; we have $126 \in RB(6)$. From [11], we know that $q^3+q^2+q+1 \in RB(q+1)$ when q is a prime power. So we have $156 \in RB(6)$. Taking k = 6, and v = 126,156 in Theorem 3 in [12], we obtain $\{756,936\} \in RB(6)$. Similarly, $4536,5616 \in RB(6)$.

COROLLARY 3.13. $151 \in B(7,25^*)$, $187 \in B(7,31^*)$, $907 \in B(7,151^*)$, $1123 \in B(7,187^*)$, $5443 \in B(7,907^*)$.

LEMMA 3.14. $\{155,162,190,938,910,918,939,952,959,981,987,988,1002,1128,1148,1149,1156,1162,1169,1191,1198,1212,1239,1197,1247,1254\} \subset R(7)$.

PROOF. The conclusion follows from Lemma 2.3, Corollary 3.13 and Table 2.

TABLE 2

v	w	a	7(v-a)+a	R(7)	check conditions
151	25	21	931	155	$130 = 7.17+7+4 \in P_5(4)$ $49 = 7(25-21)+21 \in B(7)$
151	25	14	973	162	$137 = 7.17+7+11 \in P_5(11)$ $91 = 7(25-14)+14 \in B(7)$
187	31	28	1141	190	$159 = 12.13+3 \in P_5(3)$ $7(31-28)+28 \in B(7)$
907	151	120	5629	938	$787 = 7.108+31 \in P_5(31)$ $7(151-120)+120 \in B(7)$
907	151	140	5509	918	$767 = 7.108+11 \in P_5(11)$ $7(151-140)+140 \in B(7)$
907	151	148	5461	910	$759 = 7.108+3 \in P_5(3)$ $7(151-148)+148 \in B(7)$
907	151	119	5635	939	$788 = 7.108+32 \in P_5(32)$ $7(151-119)-119 \in B(7)$
907	151	112	5677	946	$795 = 7.108+39 \in P_5(39)$ $7(151-112)+112 \in B(7)$
907	151	106	5713	952	$801 = 7.108+45 \in P_5(45)$ $7(151-106)+106 \in B(7)$
907	151	99	5755	959	$808 = 7.108+52 \in P_5(52)$ $7(151-99)+99 \in B(7)$
907	151	71	5923	987	$836 = 7.108+80 \in P_5(80)$ $7(151-71)+71 \in B(7)$
907	151	70	5929	988	$837 = 7.108+81 \in P_5(81)$ $7(151-70)+70 \in B(7)$
907	151	56	6013	1002	$851 = 7.108+95 \in P_5(95)$ $7(151-56)+56 \in B(7)$

Table 2 Continued

v	w	a	7(v-a)+a	R(7)	check conditions
1123	187	162	6889	1148	$961 = 8.117+25 \in P_5(25)$ $7(187-162)+162 \in B(7)$
1123	187	182	6769	1128	$941 = 8.117+5 \in P_5(5)$ $7(187-182)+182 \in B(7)$
1123	187	161	6895	1149	$962 = 8.117+26 \in P_5(26)$ $7(187-161)+161 \in B(7)$
1123	187	154	6937	1156	$969 = 8.117+33 \in P_5(33)$ $7(187-154)+154 \in B(7)$
1123	187	148	6973	1162	$975 = 8.117+39 \in P_5(39)$ $7(187-148)+148 \in B(7)$
1123	187	141	7015	1169	$982 = 8.117+46 \in P_5(46)$ $7(187-141)+141 \in B(7)$
1123	187	119	7147	1191	$1004 = 8.117+68 \in P_5(68)$ $7(187-119)+119 \in B(7)$
1123	187	113	7183	1197	$1010 = 8.117+74 \in P_5(74)$ $7(187-113)+113 \in B(7)$
1123	187	112	7189	1198	$1011 = 8.117+75 \in P_5(75)$ $7(187-112)+112 \in B(7)$
1123	187	98	7273	1212	$1025 = 8.117+89 \in P_5(89)$ $7(187-98)+98 \in B(7)$
1123	187	71	7435	1239	$1052 = 8.117+116 \in P_5(116)$ $7(187-71)+71 \in B(7)$
1123	187	63	7483	1247	$1060 = 7.124+68+124 \in P_5(124)$ $7(187-63)+63 \in B(7)$
1123	187	56	7525	1254	$1067 = 7.131+19+131 \in P_5(131)$ $7(187-56)+56 \in B(7)$

LEMMA 3.15. If there exists a TD(7,m), and m+7 $\in$ B(7), then, 7m+7 $\in$ B(7).

PROOF. Set k = 7 and λ = 1 in Lemma 2.10. Then the conclusion follows.

COROLLARY 3.16. {99,246,386,442,484,533,687,729,834,1023,1324, 1366,1373,1562,1618,1716,1758,1765,2003,2402,2787,3571,3865, 3914,4306,3431,3774,2696,3725} $\subset$ R(7).

PROOF. We can obtain the result from Lemma 3.15 and Table 3.

TABLE 3

m	7m+7	R(7)	check condition
84	595	99	15 $\in$ R(7)
210	1477	246	36 $\in$ R(7)
330	2317	386	56 $\in$ R(7)
378	2653	442	64 $\in$ R(7)
414	2905	484	70 $\in$ R(7)
456	3199	533	77 $\in$ R(7)
588	4123	687	99 $\in$ R(7)
624	4375	729	105 $\in$ R(7)
714	5005	834	120 $\in$ R(7)
876	6139	1023	147 $\in$ R(7)
1134	7945	1324	190 $\in$ R(7)
1170	8197	1366	196 $\in$ R(7)
1176	8239	1373	197 $\in$ R(7)
1338	9373	1562	224 $\in$ R(7)
1386	9709	1618	232 $\in$ R(7)
1470	10297	1716	246 $\in$ R(7)
1506	10549	1758	252 $\in$ R(7)
1512	10591	1765	253 $\in$ R(7)
1716	12019	2003	287 $\in$ R(7)
2058	14413	2402	344 $\in$ R(7)
2310	16177	2696	386 $\in$ R(7)
2388	16723	2787	399 $\in$ R(7)
2940	20587	3431	491 $\in$ R(7)
3060	21427	3571	511 $\in$ R(7)
3192	22351	3725	533 $\in$ R(7)
3234	22645	3774	540 $\in$ R(7)
3312	23191	3865	553 $\in$ R(7)
3354	23485	3914	560 $\in$ R(7)
3690	25837	4306	616 $\in$ R(7)

LEMMA 3.17. Suppose 7m+s, 7t+s ∈ R(7), where s = 0,1, $0 \le t \le m$, and suppose there exists a TD(9,m). Then 56m+7t+s ∈ R(7).

PROOF. Take k = n = 8 in Lemma 2.5, noting that there exists a TD(8,8) and a TD(8,7). Then we have a {8}-GDD having group-type $(7m)^8 (7t)^1$. Hence the first result is obtained. The second result is obtained by filling in groups.

COROLLARY 3.18. {449,456,463,505,512,519,561,568,953,960,967, 1009,1016,1051,1058,1289,1296,1303,1324,1345,1352,1387,1394, 1408,1443,1450,1513,1520,1527,1548,1569,1576,1611,1618,1632, 1667,1674,1702,2297,2304,2311,2332,2353,2360,2395,2451,2458, 2486,2493,2549,2584,2759,2780,2808,2843,2850,2899,2906,2934, 2976,2990,2997,3151,3172,3235,3256,3291,3298,3326,3333,3368, 3382,3424,3480,3522,3207,3221,3249,3284,3305,3340,3347,3375, 3417,3445,3473,3585,3592,3599,3620,3641,3648,3683,3690,3704, 3739,3746,3774,3781,3816,3830,3837,3872,3928,3970,3648,3655, 3676,3697,3760,3795,3802,3872,3886,3893,3984,4026,4033,4040, 4082,4089,4096,4047,4068,4131,4138,4152,4187,4194,4222,4229, 4264,4278,4285,4320,4376,4418,4425,4432,4474,4481,4488,4495, 4516,4523,4537,4103,4124,4145,4208,4243,4250,4334,4341,4530, 4544,4551,4572,4579,4593,4600,4586,4516,4670,4677,4712,4733, 4824,4866,4873,4880,4922,4929,4936,4943,4964,4971,4985,4992, 4999,5013,5020,5041,4572,4635,4642,4656,4691,4698,4726,4768, 4782,4880,4978,5027,5048,5055,5069,5076,5097,5104} ⊂ R(7).

PROOF. The conclusion follows from Table 4.

TABLE 4

m	t	7m+1	7t+1	56m+7t+1	m	t	7m+1	7t+1	56m+7t+1
8	0	57	1	449	17	8	120	57	1009
8	1	57	8	456	17	9	120	64	1016
8	2	57	15	463	17	14	120	99	1051
8	8	57	57	505	17	15	120	106	1058
9	1	64	8	512	23	0	162	1	1289
9	2	64	15	519	23	1	162	8	1296
9	8	64	57	561	23	2	162	15	1303
9	9	64	64	568	23	5	162	36	1324
17	0	120	1	953	23	8	162	57	1345
17	1	120	8	960	23	9	162	64	1352
17	2	120	15	967	23	14	162	99	1387

TABLE 4 Continued

m	t	7m+1	7t+1	56m+7t+1	m	t	7m+1	7t+1	56m+7t+1
23	15	162	106	1394	56	55	393	386	3522
23	17	162	120	1408	57	2	400	15	3207
23	22	162	155	1443	57	5	400	36	3221
23	23	162	162	1450	57	9	400	64	3249
27	0	190	1	1513	57	14	400	99	3284
27	1	190	8	1520	57	17	400	120	3305
27	2	190	15	1527	57	22	400	155	3340
27	5	190	36	1548	57	23	400	162	3347
27	8	190	57	1569	57	27	400	190	3375
27	9	190	64	1576	57	33	400	232	3417
27	14	190	99	1611	57	36	400	253	3445
27	15	190	106	1618	57	41	400	288	3473
27	17	190	120	1632	64	0	449	1	3585
27	22	190	155	1667	64	1	449	8	3592
27	23	190	162	1674	64	2	449	15	3599
27	27	190	190	1702	64	5	449	36	3620
41	0	288	1	2297	64	8	449	57	3641
41	1	288	8	2304	64	9	449	64	3648
41	2	288	15	2311	64	14	449	99	3683
41	5	288	36	2332	64	15	449	106	3690
41	8	288	57	2353	64	17	449	120	3704
41	9	288	64	2360	64	22	449	155	3739
41	14	288	99	2395	64	23	449	162	3746
41	22	288	155	2451	64	27	449	190	3774
41	23	288	162	2458	64	28	449	197	3781
41	27	288	190	2486	64	33	449	232	3816
41	28	288	197	2493	64	35	449	246	3830
41	36	288	253	2549	64	36	449	253	3837
41	41	288	288	2584	64	41	449	288	3872
49	2	344	15	2759	64	49	449	344	3928
49	5	344	36	2780	64	55	449	386	3970
49	9	344	64	2808	65	1	456	8	3648
49	14	344	99	2843	65	2	456	15	3655
49	15	344	106	2850	65	5	456	36	3676
49	22	344	155	2899	65	8	456	57	3697
49	23	344	162	2906	65	17	456	120	3760
49	27	344	190	2934	65	22	456	155	3795
49	33	344	232	2976	65	23	456	162	3802
49	35	344	246	2990	65	33	456	232	3872
49	36	344	253	2997	65	35	456	246	3886
56	2	393	15	3151	65	36	456	253	3893
56	5	393	21	3172	65	49	456	344	3984
56	14	393	99	3235	65	55	456	386	4026
56	17	393	120	3256	65	56	456	393	4033
56	22	393	155	3291	65	57	456	400	4040
56	23	393	162	3298	65	63	456	442	4082
56	27	393	190	3326	65	64	456	449	4089
56	28	393	197	3333	65	65	456	456	4096
56	33	393	232	3368	72	2	505	15	4047
56	35	393	246	3382	72	5	505	36	4068
56	41	393	288	3424	72	14	505	99	4131
56	49	393	344	3480	72	15	505	106	4138

TABLE 4 Continued

m	t	7m+1	7t+1	56m+7t+1
72	17	505	120	4152
72	22	505	155	4187
72	23	505	162	4194
72	27	505	190	4222
72	28	505	197	4229
72	33	505	232	4264
72	35	505	246	4278
72	36	505	253	4285
72	41	505	288	4320
72	49	505	344	4376
72	55	505	386	4418
72	56	505	393	4425
72	57	505	400	4432
72	63	505	442	4474
72	64	505	449	4481
72	65	505	456	4488
72	66	505	463	4495
72	69	505	484	4516
72	70	505	491	4523
72	72	505	505	4537
73	2	512	15	4103
73	5	512	36	4124
73	8	512	57	4145
73	17	512	120	4208
73	22	512	155	4243
73	23	512	162	4250
73	35	512	246	4334
73	36	512	253	4341
73	63	512	442	4530
73	65	512	456	4544
73	66	512	463	4551
73	69	512	484	4572
73	70	512	491	4579
73	72	512	505	4593
73	73	512	512	4600
80	5	561	36	4516
80	15	561	106	4586
80	27	561	190	4670
80	28	561	197	4677
80	33	561	232	4712
80	36	561	253	4733
80	49	561	344	4824
80	55	561	386	4866
80	56	561	393	4873
80	57	561	400	4880
80	63	561	442	4922
80	64	561	449	4929
80	65	561	456	4936
80	66	561	463	4943
80	69	561	484	4964
80	70	561	491	4971
80	72	561	505	4985
80	73	561	512	4992
80	74	561	519	4999
80	76	561	533	5013
80	77	561	540	5020
80	80	561	561	5041
81	5	568	36	4572
81	14	568	99	4635
81	15	568	106	4642
81	17	568	120	4656
81	22	568	155	4691
81	23	568	162	4698
81	27	568	190	4726
81	33	568	232	4768
81	35	568	246	4782
81	49	568	344	4880
81	63	568	442	4978
81	70	568	491	5027
81	73	568	512	5048
81	74	568	519	5055
81	76	568	533	5069
81	77	568	540	5076
81	80	568	561	5097
81	81	568	568	5104

COROLLARY 3.19. {476,504,644,672,686,693,1792,1820,1848,1862, 1869,1897,1939,1988,20162324;2352,2366,2373,2401,2443,2492, 2520,2548,2583,3164,3206,3213,3241,3283,3332,3360,3388,3423, 3528,3192,3220,3248,3262,3269,3297,3339,3416,3444,3479,3584, 3591,3612,3640,3654,3661,3689,3731,3780,3808,3836,3871,3976, 3983,4025,4032,4060,4088,4102,4109,4137,4179,4228,4256,4284, 4319,4424,4431,4473,4480,4508,4522,4536,4116,4144,4158,4165, 4193,4235,4312,4340,4375,4487,4529,4557,4571,4585,4592,4452,

4494,4501,4620,4648,4676,4711,4816,4823,4865,4872,4900,4914, 4928,4935,4963,4977,4550,4627,4704,4732,4767,4879,4921,4956, 4970,4984,4991,5019,5033,5040,4998,5005,5075,5124,5152,5180, 5215,5320,5327,5369,5376,5404,5418,5432,5439,5467,5481,5488, 5537,5544,5096,5166,5173,5201,5243,5292,5348,5383,5495,5586, 5607,5635,5649,5656,5705,5712,5446,5453,5523,5572,5600,5628, 5663,5768,5775,5817,5824,5852,5866,5880,5887,5915,5929,5936, 5985,5999,6027,6055,5614,5621,5649,5691,5740,5796,5831,5943, 5992,6006,6020,6034,6041,6069,6083,6090,6139,6146,6174,6202, 6216,6223,5908,5950,5957,6076,6104,6132,6167, 6272,6279,6321, 6328,6356,6370,6384,6391,6419,6433,6440,6489,6496,6524,6552, 6566,6573,6615} $\subset$ R(7).

PROOF. The solution follows from Table 5.

TABLE 5

m	t	7m	7t	56m+7t	m	t	7m	7t	56m+7t
8	4	56	28	476	56	21	392	147	3283
8	8	56	56	504	56	28	392	196	3332
11	4	77	28	644	56	32	392	224	3360
11	8	77	56	672	56	36	392	252	3388
11	10	77	70	686	56	41	392	287	3423
11	11	77	77	693	56	56	392	392	3528
32	0	224	0	1792	57	0	399	0	3192
32	4	224	28	1820	57	4	399	28	3220
32	8	224	56	1848	57	8	399	56	3248
32	10	224	70	1862	57	10	399	70	3262
32	11	224	77	1869	57	11	399	77	3269
32	15	224	105	1897	57	15	399	105	3297
32	21	224	147	1939	57	21	399	147	3339
32	28	224	196	1988	57	32	399	224	3416
32	32	224	224	2016	57	36	399	252	3444
41	4	287	28	2324	57	41	399	287	3479
41	8	287	56	2352	57	56	399	392	3584
41	10	287	70	2366	57	57	399	399	3591
41	11	287	77	2373	64	4	448	28	3612
41	15	287	105	2401	64	8	448	56	3640
41	21	287	147	2443	64	10	448	70	3654
41	28	287	196	2492	64	11	448	77	3661
41	32	287	224	2520	64	15	448	105	3689
41	36	287	252	2548	64	21	448	147	3731
41	41	287	287	2583	64	28	448	196	3780
56	4	392	28	3164	64	32	448	224	3808
56	10	392	70	3206	64	36	448	252	3836
56	11	392	77	3213	64	41	448	287	3871
56	15	392	105	3241	64	56	448	392	3976

TABLE 5 Continued

m	t	7m	7t	56m+7t	m	t	7m	7t	56m+7t
64	57	448	399	3983	79	73	553	511	4935
64	63	448	441	4025	79	77	553	539	4963
64	64	448	448	4032	79	79	553	553	4977
72	4	504	28	4060	80	10	560	70	4550
72	8	504	56	4088	80	21	560	147	4627
72	10	504	70	4102	80	32	560	224	4704
72	11	504	77	4109	80	36	560	252	4732
72	15	504	105	4137	80	41	560	287	4767
72	21	504	147	4179	80	57	560	399	4879
72	28	504	196	4228	80	63	560	441	4921
72	32	504	224	4256	80	68	560	476	4956
72	36	504	252	4284	80	70	560	490	4970
72	41	504	287	4319	80	72	560	504	4984
72	56	504	392	4424	80	73	560	511	4991
72	57	504	399	4431	80	77	560	539	5019
72	63	504	441	4473	80	79	560	553	5033
72	64	504	448	4480	80	80	560	560	5040
72	68	504	476	4508	88	10	616	70	4998
72	70	504	490	4522	88	11	616	77	5005
72	72	504	504	4536	88	21	616	147	5075
73	4	511	28	4116	88	28	616	196	5124
73	8	511	56	4144	88	32	616	224	5152
73	10	511	70	4158	88	36	616	252	5180
73	11	511	77	4165	88	41	616	287	5215
73	15	511	105	4193	88	56	616	392	5320
73	21	511	147	4235	88	57	616	399	5327
73	32	511	224	4312	88	63	616	441	5369
73	36	511	252	4340	88	64	616	448	5376
73	41	511	287	4375	88	68	616	476	5404
73	57	511	399	4487	88	70	616	490	5418
73	64	511	448	4529	88	72	616	504	5432
73	68	511	476	4557	88	73	616	511	5439
73	70	511	490	4571	88	77	616	539	5467
73	72	511	504	4585	88	79	616	553	5481
73	73	511	511	4592	88	80	616	560	5488
79	4	553	28	4452	88	87	616	609	5537
79	10	553	70	4494	88	88	616	616	5544
79	11	553	77	4501	91	0	637	0	5096
79	28	553	196	4620	91	10	637	70	5166
79	32	553	224	4648	91	11	637	77	5173
79	36	553	252	4676	91	15	637	105	5201
79	41	553	287	4711	91	21	637	147	5243
79	56	553	392	4816	91	28	637	196	5292
79	57	553	399	4823	91	36	637	252	5348
79	63	553	441	4865	91	41	637	287	5383
79	64	553	448	4872	91	57	637	399	5495
79	68	553	476	4900	91	70	637	490	5586
79	70	553	490	4914	91	73	637	511	5607
79	72	553	504	4928	91	77	637	539	5635

TABLE 5 Continued

m	t	7m	7t	56m+7t	m	t	7m	7t	56m+7t
91	79	637	553	5649	99	72	693	504	6034
91	80	637	560	5656	99	73	693	511	6041
91	87	637	609	5705	99	77	693	539	6069
91	88	637	616	5712	99	79	693	553	6083
96	10	672	70	5446	99	80	693	560	6090
96	11	672	77	5453	99	87	693	609	6139
96	21	672	147	5523	99	88	693	616	6146
96	28	672	196	5572	99	92	693	644	6174
96	32	672	224	5600	99	96	693	672	6202
96	36	672	252	5628	99	98	693	689	6216
96	41	672	287	5663	99	99	693	693	6223
96	56	672	392	5768	105	4	735	28	5908
96	57	672	399	5775	105	10	735	70	5950
96	63	672	441	5817	105	11	735	77	5957
96	64	672	448	5824	105	28	735	196	6076
96	68	672	476	5852	105	32	735	224	6104
96	70	672	490	5866	105	36	735	252	6132
96	72	672	504	5880	105	41	735	287	6167
96	73	672	511	5887	105	56	735	392	6272
96	77	672	539	5915	105	57	735	399	6279
96	79	672	553	5929	105	63	735	441	6321
96	80	672	560	5936	105	64	735	448	6328
96	87	672	609	5985	105	68	735	476	6356
96	88	672	616	5999	105	70	735	490	6370
96	92	672	644	6027	105	72	735	504	6384
96	96	672	672	6055	105	73	735	511	6391
99	10	693	70	5614	105	77	735	539	6419
99	11	693	77	5621	105	79	735	553	6433
99	15	693	105	5649	105	80	735	560	6440
99	21	693	147	5691	105	87	735	609	6489
99	28	693	196	5740	105	88	735	616	6496
99	36	693	252	5796	105	92	735	644	6524
99	41	693	287	5831	105	96	735	672	6552
99	57	693	399	5943	105	98	735	686	6566
99	64	693	448	5992	105	99	735	693	6573
99	68	693	476	6006	105	105	735	735	6615
99	70	693	490	6020					

LEMMA 3.20. If there exists a TD(7,s-1) and $r,s \in R(7)$, then

$$(s-1)r+1 \in R(7).$$

PROOF. Take $k = d = 7$, $v = 6r+1$ in Lemma 2.1. Then $m+7 \in F_{\{7\}}[7]$ is equivalent to $m+7 \in B(7)$. So we can let $m+7 = 6s+1$, $m/6 = s-1$. Then we have our conclusion.

COROLLARY 3.21. {946,833,785,609,826,1036,1135,1141,1331,1471, 1541,1569,1576,1561,1765,1786,1933,1981,2017,2129,2269,2485, 2737,2913,2941,3081,3095,3137,3193,3144,3242,3389,3438,3521, 3529,3536,3585,3634,3732,3745,3851,3872,3921,3934,3977,4236, 4257,4264,4313,4333,4411,4417,4803,4810,4831,4852,4859,5446} $\subset$ R(7).

PROOF. The conclusion follows from Lemma 3.20, Lemma 3.1, Corollary 3.3, Corollary 3.5, Lemma 3.9, Corollary 3.11, Lemma 3.14, Corollary 3.16, Corollary 3.18 and Table 6.

TABLE 6

(s-1)r+1	(s-1)r+1
1135 = (8-1)·162+1	4257 = (77-1)·56+1
1981 = (56-1)·36+1	4333 = (77-1)·57+1
826 = (56-1)·15+1	833 = (105-1)·8+1
1541 = (56-1)·28+1	1561 = (105-1)·15+1
3081 = (56-1)·56+1	2913 = (105-1)·28+1
3521 = (56-1)·64+1	3745 = (105-1)·36+1
3851 = (56-1)·70+1	1576 = (106-1)·15+1
4236 = (56-1)·77+1	2941 = (106-1)·28+1
1569 = (57-1)·28+1	785 = (99-1)· 8+1
2017 = (57-1)·36+1	1471 = (99-1)·15+1
3137 = (57-1)·56+1	3095 = (8-1)·442+1
3193 = (57-1)·57+1	3144 = (8-1)·449+1
3921 = (57-1)·70+1	3242 = (8-1)·463+1
4313 = (57-1)·77+1	3389 = (8-1)·484+1
946 = (64-1)·15+1	3438 = (8-1)·491+1
1765 = (64-1)·28+1	3536 = (8-1)·505+1
3529 = (64-1)·56+1	3585 = (8-1)·512+1
2269 = (64-1)·36+1	3634 = (8-1)·519+1
4411 = (64-1)·70+1	3732 = (8-1)·533+1
4852 = (64-1)·77+1	1331 = (8-1)·190+1
1036 = (70-1)·15+1	3872 = (8-1)·553+1
1933 = (70-1)·28+1	3977 = (8-1)·568+1
2485 = (70-1)·36+1	4264 = (8-1)·609+1
3934 = (70-1)·57+1	4803 = (8-1)·686+1
4417 = (70-1)·64+1	4810 = (8-1)·687+1
4831 = (70-1)·70+1	4859 = (8-1)·694+1
609 = (77-1)· 8+1	1786 = (120-1)·15+1
1141 = (77-1)·15+1	
2129 = (77-1)·28+1	
2737 = (77-1)·36+1	

LEMMA 3.22 ([8] Theorem 9.1). If $m+d \in F_{\{k\}}[d]$ and there exists a TD(k,m), then for every $v \in B(k)$, $mv+d \in B(k)$.

COROLLARY 3.23. If $m+d \in B(7)$, d = 0,1,7, and there exists a TD(7,m), then for every $v \in B(7)$, $(mv+d-1)/6 \in R(7)$.

PROOF. Set k = 7, d = 0,1,7 in Lemma 3.22, then the conclusion follows.

COROLLARY 3.24. {441,840,1085,1086,1079,1127,1134,539,1275, 1380,1365,1618,1624,1625,1722,1723,2738,2745,2751,2752,2696, 2702,2703,2794,3088,3094,3095,3130,3136,3186,3192,3382,3430, 3438,3773,3781,4655,5006,5111,5096,5097,5145,5194} $\subset$ R(7).

PROOF. The conclusion follows from Corollary 3.23 and Table 7.

TABLE 7

R(7)	v	m	d	check condition
441	7	378	7	64 $\in$ R(7)
840	7	720	1	120 $\in$ R(7)
1086	7	931	0	155 $\in$ R(7)
1085	7	930	1	
1079	7	924	7	
1134	7	972	1	162 $\in$ R(7)
1127	7	966	7	
1625	7	1393	0	232 $\in$ R(7)
1624	7	1392	1	
1618	7	1386	7	
539	7	462	1	77 $\in$ R(7)
1723	7	1477	0	246 $\in$ R(7)
1722	7	1476	1	
2703	7	2317	0	386 $\in$ R(7)
2702	7	2316	1	
2696	7	2310	7	386 $\in$ R(7)
2752	7	2359	0	393 $\in$ R(7)
2751	7	2358	1	
2745	7	2352	7	
2738	7	2346	7	392 $\in$ R(7)
3130	7	2682	7	448 $\in$ R(7)
3136	7	2688	1	
3192	7	2736	1	456 $\in$ R(7)
3186	7	2730	7	
2794	7	2394	7	400 $\in$ R(7)
3095	7	2653	0	442 $\in$ R(7)
3094	7	2652	1	

TABLE 7 Continued

R(7)	v	m	d	check condition
3088	7	2646	7	
3438	49	421	0	70 ∈ R(7)
3382	49	414	7	
3430	49	420	1	
3781	49	463	0	77 ∈ R(7)
3773	49	462	1	
5194	49	636	1	106 ∈ R(7)
5145	49	630	1	105 ∈ R(7)
5097	49	624	7	
5006	91	330	7	56 ∈ R(7)
5111	91	337	0	
5096	91	336	1	
1275	91	84	7	15 ∈ R(7)
1380	91	91	0	
1365	91	90	1	

LEMMA 3.25. If there exists a TD(8,m), then 63m ∈ B(8,7m).

PROOF. Take n = 8 in Lemma 2.6.

COROLLARY 3.26. {4032,6237,6615} ∈ R(7).

PROOF. Set m = 64,99,105 in Lemma 3.25. From Corollary 3.11 and Corollary 3.19, we know 448,735,693 ∈ R(7). Then the conclusion follows.

LEMMA 3.27. {5447,5454,5482,5488,5489,5516,5524,5544,5545,5538, 5579,5636,5684,5678,5685,5691,5692,5733,5734,5740,5741,5748, 5755,5776,5782} ⊂ R(7).

PROOF. The conclusion follows from Lemma 2.3, Corollary 3.13 and Table 8.

TABLE 8

v	w	a	R(7)	check condition
5443	907	903	5447	4540 = 8.567+4 ∈ $P_5(4)$ 7(907-903)+903 ∈ B(7)

TABLE 8 Continued

v	w	a	R(7)	check condition
5443	907	896	5454	$4547 = 8.567+11 \in P_5(11)$ $7(907-896)+896 \in B(7)$
5443	907	868	5482	$4575 = 8.567+39 \in P_5(39)$ $7(907-868)+868 \in B(7)$
5443	907	862	5488	$4581 = 8.567+45 \in P_5(45)$ $7(907-862)+862 \in B(7)$
5443	907	861	5489	$4582 = 8.567+46 \in P_5(46)$ $7(907-861)+861 \in B(7)$
5443	907	834	5516	$4609 = 8.567+73 \in P_5(73)$ $7(907-834)+834 \in B(7)$
5443	907	826	5524	$4617 = 8.567+81 \in P_5(81)$ $7(907-826)+826 \in B(7)$
5443	907	806	5544	$4637 = 8.567+101 \in P_5(101)$ $7(907-806)+806 \in B(7)$
5443	907	805	5545	$4638 = 8.567+102 \in P_5(102)$ $7(907-805)+805 \in B(7)$
5443	907	812	5538	$4631 = 8.567 + 95 \in P_5(95)$ $7(907-812)+812 \in B(7)$
5443	907	771	5579	$4672 = 8.567+136 \in P_5(136)$ $7(907-771)+771 \in B(7)$
5443	907	714	5636	$4729 = 8.567+193 \in P_5(193)$ $7(907-714)+714 \in B(7)$
5443	907	672	5678	$4771 = 8.567+235 \in P_5(235)$ $7(907-672)+672 \in B(7)$
5443	907	666	5684	$4777 = 8.567+241 \in P_5(241)$ $7(907-666)+666 \in B(7)$
5443	907	659	5691	$4784 = 8.567+248 \in P_5(248)$ $7(907-659)+659 \in B(7)$

TABLE 8 Continued

v	w	a	R(7)	check condition
5443	907	665	5685	$4778 = 8.567+242 \in P_5(242)$ $7(907-665)+665 \in B(7)$
5443	907	658	5692	$4785 = 8.567+249 \in P_5(249)$ $7(907-658)+658 \in B(7)$
5443	907	617	5733	$4826 = 8.567+290 \in P_5(290)$ $7(907-617)+617 \in B(7)$
5443	907	616	5734	$4827 = 8.567+291 \in P_5(291)$ $7(907-616)+616 \in B(7)$
5443	907	610	5740	$4833 = 8.567+297 \in P_5(297)$ $7(907-610)+610 \in B(7)$
5443	907	609	5741	$4834 = 8.567+298 \in P_5(298)$ $7(907-609)+609 \in B(7)$
5443	907	602	5748	$4841 = 8.567+305 \in P_5(305)$ $7(907-602)+602 \in B(7)$
5443	907	595	5755	$4848 = 8.567+312 \in P_5(312)$ $7(907-595)+595 \in B(7)$
5443	907	574	5776	$4869 = 8.567+333 \in P_5(333)$ $7(907-574)+574 \in B(7)$
5443	907	568	5782	$4875 = 8.567+339 \in P_5(339)$ $7(907-568)+568 \in B(7)$

COROLLARY 3.28. $\{1155,1050,2772,5390\} \subset R(7)$.

PROOF. Set $s = 15$, $r = 70,77$; $r = 77$, $s = 36,70$ and $k = 7$ in Lemma 3.7, noticing that there exist TD(7,70), TD(7,77). We have the result.

4. CONCLUSIONS

We have obtained some designs above. In this part we will give two lemmas in order to prove Theorem 1.1 and then give the proof of that theorem.

LEMMA 4.1. If there exist integers $e_{i+1} = e_i+1$, $i = 0,1,\ldots,7$, and an integer A ($\geq \max\{780,e_7\}$) so that $7e_i+1 \in R(7)$, and when $A \leq r \leq 8A+e_7$, $7r+1 \in R(7)$, then, for every $r \geq A$, $7r+1 \in R(7)$.

PROOF. We use induction on r. We know that $7r+1 \in R(7)$ when $A \leq r \leq 8A+e_7$. Assume that $7r+1 \in R(7)$ when $A \leq r \leq N-1$; we want to prove $7N+1 \in R(7)$. If $N \leq 8A+e_7$, then $7N+1 \in R(7)$; we can assume $N \geq 8A+e_7+1$, because N can be expressed as $8s+e_i$, $i = 0,1,\ldots,7$, $7N+1 = 56s+7e_i+1$. From $N \geq 8A+e_7+1$ we know $s \geq A+1 \geq e_7+1 > e_i$, and that $s \geq 781$ (TD(9,s)). Obviously, $A \leq s \leq N-1$, so $7s+1 \in R(7)$. From Lemma 3.17, $7N+1 \in R(7)$.

By induction we see that $7r+1 \in R(7)$ is true for every $r \geq A$.

By a similar method, we can obtain:

LEMMA 4.2. If there exist integers $e_{i+1} = e_i+1$, $i = 0,1,\ldots,7$, and an integer A ($\geq \max\{780,e_7\}$) so that $7e_i \in R(7)$, and if $A \leq r \leq 8A+e_7$, $7r \in R(7)$, then for every $r \geq A$, $7r \in R(7)$.

LEMMA 4.3. $E \subset U(7)$, $F \subset V(7)$, where

E = {0,1,2,5,8,9,14,15,17,22,23,27,28,33,35,36,41,49,55,56, 57,63,64,65,66,69,70,72,73,74,76,77,80,81,98,99,104,105,106,108, 112,119,120,129,131,134,135,136,137,138,140,141,143,144,145,146, 147,150,151,153,154,155,161,162,164,165,170,171,173,178,179,182, 184,185,186,189,190,192,193,195,196,197,198,199,201,206,207,210, 216,217,218,220,221,223,224,225,230,231,232,233,238,239,243,245, 246,251,252,255,265,273,276,286,287,288,289,304,324,328,329,330, 333,336,337,342,343,344,345,350,351,353,355,356,361,364,369,385, 386,391,392,393,394,397,398,400,401,406,407,409,414,415,416,417, 419,420,425,427,428,433,440,441,442,447,448,449,450,453,455,456, 457,458,460,462,463,464,465,469,470,471,472,473,475,476,477,478, 481,482,483,484,488,489,490,491,492,496,497,503,504,505,510,511,

512,513,514,517,519,520,521,522,525,526,527,528,529,532,533,534, 535,537,539,540,542,543,546,547,548,550,552,553,555,556,559,560, 561,567,568,569,575,576,577,578,581,583,584,585,586,589,590,591, 592,593,598,599,601,603,604,605,606,607,608,609,611,612,615,616, 617,619,620,625,630,631,632,633,639,640,641,642,645,646,647,648, 649,650,653,654,655,656,657,662,663,665,667,668,670,671,673,675, 676,681,683,684,685,687,689,690,693,694,695,696,697,703,704,705, 706,709,710,711,712,713,714,715,716,717,718,720,721,722,724,725, 728,729,730},

F = {4,8,10,11,15,21,28,32,36,41,56,57,63,64,70,72,73,77, 79,80,87,88,91,92,93,96,98,99,105,106,110,118,119,120,130,134, 141,147,148,150,155,161,162,163,164,165,166,167,168,171,177,190, 195,196,197,223,224,232,246,256,260,264,266,267,271,277,283,284, 288,328,332,336,338,339,343,349,355,356,360,364,369,386,391,392, 393,397,400,442,448,452,456,458,459,460,463,464,466,467,469,471, 476,477,480,484,488,489,490,492,497,503,504,512,513,516,520,522, 523,527,533,535,539,540,544,548,553,562,568,569,575,576,580,584, 586,587,588,591,592,594,595,597,599,604,605,608,612,616,617,619, 620,625,631,632,633,637,639,640,641,642,643,644,646,647,648,650, 651,653,655,656,660,661,664,665,668,672,673,676,681,688,689,695, 696,697,700,702,703,704,705,708,709,710,711,712,713,714,715,717, 719,720,725,728,732,735,736,738,740,742,745,749,756,760,761,764, 767,768,769,770,772,774,776,777,778,779,781,783,784,785,789,791, 792,796,798,800,801,802,803,804,805,807,808,809,813,815,816,819, 820,824,825,828,831,832,833,836,838,840,841,844,845,847,848,849, 850,851,855,856,857,858,860,861,862,863,865,867,868,869,870,872, 876,877,878,881,882,886,889,896,897,903,904,908,910,912,913,917, 919,920,927,928,932,936,938,945}.

PROOF. This can be obtained from all the Lemmas of part 3.

LEMMA 4.4. (1) $r \in V(7)$, where $8183 \leq r \leq 70000$.
(2) $r \in U(7)$, where $4134 \leq r \leq 35000$.

PROOF. (1) Take each $m \geq 120$ from F for which there exists a TD(9,m) in Lemma 3.17 and take each $t \leq m$ from F in Lemma 3.17

for given m; then we can obtain $56m+7t \in R(7)$, i.e. $8m+t \in V(7)$. Let F_1 denote the set of obtained values $8m+t$ and let $F' = F_1 \cup F$, then take each m from F_1 (there exists a TD(9,m) for such m) and take each $t \le m$ from F' for each m, then we also can obtain $8m+t \in V(7)$ for such m, t, and the conclusion follows. We do this by computer.

(2) Similar to the above.

LEMMA 4.5. (1) $r \in V(7)$, where $r \ge 8183$.
(2) $r \in U(7)$, where $r \ge 4134$.

PROOF. (1) Take $A = 8183$, $e_i = 708+i$, $i = 0,1,\ldots,7$, in Lemma 4.2, then (1) follows from Lemma 4.3 and Lemma 4.4.

(2) Take $A = 4134$, $e_i = 709+i$, $i = 0,1,\ldots,7$, in Lemma 4.1, then (2) follows from Lemma 4.3 and Lemma 4.4.

PROOF OF THEOREM 1.1. This is now a direct consequence of Lemma 4.5.

REFERENCES

[1] R.C. Mullin and D.R. Stinson, Pairwise balanced designs with block sizes 6t+1, Graphs and Combinatorics 3(1987), 365-377.

[2] A.E. Brouwer, The number of mutually orthogonal latin squares -- a table up to order 100,000, Math. Centr. Report ZW123, June, 1979.

[3] H. Hanani, Balanced incomplete block designs and related designs, Discrete Math. 11(1975), 255-369.

[4] A.E. Brouwer, G.H.J. Van Rees, More mutually orthogonal latin squares, Discrete Math. 39(1982), 263-281.

[5] R.M. Wilson, Concerning the number of mutually orthogonal latin squares, Discrete Math. 9(1974), 181-198.

[6] H.F. MacNeish, Euler squares, Ann. Math. 23(1922), 221-227.

[7] H. Hanani, The existence and construction of balanced incomplete block designs, Ann. Math. Statist. 32(1961), 361-386.

[8] R.M. Wilson, An existence theory for pairwise balanced designs, 1. Composition theorems and morphisms, J. Combinatorial Theory (A) 13(1972), 220-245.

[9] R. Mathon, A. Rosa, Tables of parameters of BIBDs with $r \leq 41$ including existence, enumeration, and resolvability results, Ann. Discrete Math. 26(1985), 275-308.

[10] R.C. Bose, On the application of finite protective geometry for deriving a certain series of balanced Kirkman arrangements, Calcutta Math. Soc. Golden Jubilee Vol., 1959, 341-354.

[11] R. Mathon, On the existence of doubly resolvable Kirkman systems and equidistant permutation arrays, Discrete Math. 30(1980), 157-172.

[12] H. Hanani, D.K. Ray-Chaudhuri and R.M. Wilson, On resolvable designs, Discrete Math. 3(1972), 343-357.

[13] Du Beiliang, On the existence of balanced incomplete block designs with $k = 8$ and $\lambda = 1$, In the Third Conf. on Combinatorics in China, Suzhou.

[14] Zhang Yusen, Yang Zhao, Four balanced incomplete block designs with $k = 7$ and $\lambda = 1$ (this volume).

Four Balanced Incomplete Block Designs with k = 7 and $\lambda = 1$

ZHANG Yusen, Dalian University of Technology, Dalian, China

YANG Zhao, Dalian University of Technology, Dalian, China

H. Hanani [1] has found that a necessary and sufficient condition for the existence of a balanced incomplete block design (BIBD) with k = 7 and $\lambda \equiv 0, 6, 7, 12, 18, 24, 30, 35, 36 \pmod{42}$ or with k = 7 and $\lambda > 30$, λ is not divisible by 2 or 3, and that

$$\lambda(v-1) \equiv 0 \pmod{6} \text{ and } \lambda v(v-1) \equiv 0 \pmod{42}.$$

For BIBD with k = 7 and $\lambda = 1$, up to now, we know only that when v = 49,91,169,217,337,421,463,883, 1723,3067 and 3319, a (v,7,1)-BIBD exists, and that a (43,7,1)-BIBD does not exist [1],[3],[4]. In this paper we give direct constructions for four new values of v, namely v = 3823,3907,4621 and 4957. We can also obtain constructions for v = 337,421,883, 1723,3067 and 3319. These constructions were found by computer.

1. DEFINITION AND NOTATIONS

It is well known that a balanced incomplete block design, denoted by (v,k,λ)-BIBD, is a pair (X,A), where X is a v-set and

A is a collection of b k-subsets of X called blocks such that each element of X is contained in exactly r blocks and any 2-subset of X is contained in exactly λ blocks. Trivially, necessary conditions for the existence of a (v,k,λ)-BIBD are

$$\lambda(v-1) \equiv 0 \pmod{k-1}$$
$$\lambda v(v-1) \equiv 0 \pmod{k(k-1)}.$$

When $k = 7$ and $\lambda = 1$, these necessary conditions are equivalent to

$$v \equiv 1 \text{ or } 7 \pmod{42}.$$

We consider the case where $v \equiv 1 \pmod{42}$ and v is a prime.

If B is a set on which subtraction is defined, then $\Delta B = \{a-b: a,b \in B, a \neq b\}$, with the multiplicities of entries retained, and if S is a family $B_1, B_2, \ldots, B_n$ of sets then $\Delta S = \cup \Delta B_i$, with multiple occurrences again counted multiply. A family S of subsets of an additive group G of order v will be called a (v,k,λ)-difference family in G if and only if all subsets which occur in S have size k and $\Delta S = \lambda(G-\{0\})$.

We use the notations in [1]:

Z(n) denotes the cyclic group of residues mod n.

Z(p,x) (only when p is a prime) denotes Z(p) with the additional information that x is the primitive root used. When X = Z(p,x), the points of x are denoted by exponents of x, so the symbol a denotes the point x^a. The brackets $\langle\ \rangle$ are used exclusively to denote blocks.

2. RESULTS

The following two lemmas come from [2].

Let G be a finite additive group. Let S be a family of subsets of G, $S = \{B_i: i \in I\}$, and define dev $S = \sum_{i \in I}(B_i+g)$ where $B_i+g = \{b+g: b \in B_i\}$. For a subset $B \subset G$, let $\Delta B = \{a-b: a,b \in B, a \neq b\}$, $\Delta S = \sum_{i\in I}\Delta B_i$.

LEMMA 1. Let S be a family of subsets of G. Then dev S is pairwise balanced with index λ if and only if $\Delta S = \lambda(G-\{0\})$.

LEMMA 2. If S is a (v,k,λ)-difference family in G, then (X,dev

S) is a (v,k,λ)-BIBD.

Suppose $v \equiv 1 \pmod{42}$ and v is a prime. Using a computer and the above lemmas, we obtain that a (v,7,1)-BIBD exists when v = 337,421,463,883,1723,3067,3319,3823,3907,4621 and 4957, $v \leq 5000$. For each BIBD the correspondent base blocks [3] are given in the Table.

REFERENCES

[1] H. Hanani, Balanced incomplete block designs and related designs, Discrete Math. 11(1975), 255-369.

[2] R. M. Wilson, Cyclotomy and difference families in elementary abelian groups, J. Number Theory 4(1972), 17-47.

[3] M. Hall Jr., Combinatorial Theory, Blaisdell, Waltham, Mass. 1967.

[4] R. Mathon, A. Rosa, Tables of parameters of BIBDs with $r \leq 41$ including existence, enumeration, and resolvability results, Ann. Discrete Math. 26(1985), 275-308.

TABLE

v	(v,y,1)-BIBD
337	X = Z(337,3) A = ⟨3a,3a+60,3a+96,3a+144,3a+192,3a+240, 3a+288⟩ mod 337 a = 0,1,...,7
421	X = Z(421,2) A = ⟨3a,3a+60,3a+120,3a+180,3a+240,3a+300, 3a+360⟩ mod 421 a = 0,1,...,9
463	X = Z(463,3) A = ⟨3a,3a+66,3a+132,3a+198,3a+264,3a+330, 3a+396⟩ mod 463 a = 0,1,...,10
883	X = Z(883,2) A = ⟨3a,3a+126,3a+252,3a+378,3a+504,3a+630, 3a+756⟩ mod 883 a = 0,1,...,20
1723	X = Z(1723,3) A = ⟨3a,3a+246,3a+492,3a+738,3a+984,3a+1230, 3a+1476⟩ mod 1723 a = 0,1,...,40
3067	X = Z(3067,2) A = ⟨3a,3a+438,3a+876,3a+1314,3a+1752,3a+2190, 3a+2628⟩ mod 3067 a = 0,1,...,72
3319	X = Z(3319,6) A = ⟨3a,3a+474,3a+948,3a+1422,3a+1896,3a+2370, 3a+2844⟩ mod 3319 a = 0,1,...,78
3823	X = Z(3823,3) A = ⟨3a,3a+546,3a+1092,3a+1638,3a+2184,3a+2730, 3a+3276⟩ mod 3823 a = 0,1,...,90
3907	X = Z(3907,2) A = ⟨3a,3a+558,3a+1116,3a+1674,3a+2232,3a+2790, 3a+3348⟩ mod 3907 a = 0,1,...,92
4621	X = Z(4621,2) A = ⟨3a,3a+660,3a+1320,3a+1980,3a+2640,3a+3300, 3a+3960⟩ mod 4621 a = 0,1,...,109
4957	X = Z(4957,2) A = ⟨3a,3a+708,3a+1416,3a+2124,3a+2832,3a+3540, 3a+4248⟩ mod 4957 a = 0,1,...,117

Index